ACPL ITEM
DISCARDED

3 1833 00301 0920

S0-BWU-891

Ch 7

THE EVOLUTION OF CAPITALISM

THE EVOLUTION OF CAPITALISM

Advisory Editor
LEONARD SILK
Editorial Board,
The New York Times

Research Associate
MARK SILK

THE MANUFACTURING POPULATION OF ENGLAND,

ITS MORAL, SOCIAL, AND PHYSICAL CONDITIONS, AND THE CHANGES WHICH HAVE ARISEN FROM THE USE OF STEAM MACHINERY;

WITH AN

Examination of Infant Labour

P[eter] Gaskell

ARNO PRESS
A NEW YORK TIMES COMPANY
New York • 1972

Reprint Edition 1972 by Arno Press Inc.

Reprinted from a copy in
The University of Illinois Library

The Evolution of Capitalism
ISBN for complete set: 0-405-04110-1
See last pages of this volume for titles.

Manufactured in the United States of America

- - - - - - - - - - - - -

Library of Congress Cataloging in Publication Data

Gaskell, P.
The manufacturing population of England.

(The Evolution of capitalism)
1836 and 1968 editions published under title: Artisans and machinery.
Reprint of the 1833 ed.
1. Labor and laboring classes--Gt. Brit. 2. Factory system--Gt. Brit. 3. Children--Employment --Gt. Brit. 4. Machinery in industry. I. Title. II. Series.
HD2356.G7G4 1972 301.44'42'0942 73-38266
ISBN 0-405-04120-9

THE MANUFACTURING POPULATION OF ENGLAND,

1731991

ITS MORAL, SOCIAL, AND PHYSICAL CONDITIONS, AND THE CHANGES WHICH HAVE ARISEN FROM THE USE OF STEAM MACHINERY;

WITH AN

Examination of Infant Labour.

BY

P. GASKELL, Esq.

"FIAT JUSTITIA."

LONDON:
BALDWIN AND CRADOCK.
PATERNOSTER-ROW.

1833.

LONDON:
BAYLIS AND LEIGHTON, JOHNSON'S COURT, FLEET-STREET.

CONTENTS.

Page

CHAPTER I.

CHARACTER AND HABITS OF THE EARLY MANUFACTURERS.

CHAPTER II.

INFLUENCE OF TEMPERATURE AND MANNERS UPON PHYSICAL DEVELOPMENT, &c. AND UPON MORALS.

CHAPTER III.

THE INFLUENCE OF THE SEPARATION OF FAMILIES, &c. &c. UPON THE MORAL AND SOCIAL CONDITION OF THE MANUFACTURING POPULATION.

Page

CHAPTER VII.

INFANT LABOUR.

CHAPTER VIII.

HEALTH—RATE OF MORTALITY, &c. &c.

CHAPTER IX.

FACTORY LABOUR IN GENERAL—PECULIAR DISEASES, &c.

CHAPTER X.

EDUCATION—RELIGION—CRIME—PAUPERISM.

Page

CHAPTER XI.

COMBINATIONS—INFLUENCE OF MACHINERY ON LABOUR.

CHAPTER XII.

TRUCK AND COTTAGE SYSTEMS.

PREFACE.

THE entire absence of a work like the present, has long been felt as a serious evil. Nothing whatever in the shape of information is to be found, except a few pamphlets, the productions of party warmth, or of a school of political economy, the dogmas of which are open to insuperable objections; whilst the vast importance of the subject itself, and the numerous interests connected with it, render it highly desirable that it should be carefully examined and dispassionately considered,—and that a book should be the result to which all parties may look with confidence, and that which the general reader may turn to for accurate information.

It is very true that committees of the House of Commons, have at different periods examined into some of the points of this inquiry.

It has however unfortunately happened that the evidence given before them, and the reports founded upon that evidence, are looked at suspiciously, and have in effect no weight or consideration. This has arisen, partly from the peculiar nature of the evidence, which too often is tinged so strongly with party spirit, that it subjects itself at once to doubt or total unbelief; and partly to the want of knowledge amongst the members of the committee, which prevented them applying such questions and trains of reasoning and deductions, as the actual condition of the subject demanded. Hence, though many valuable facts have been elicited by these inquiries, they are in such a crude form, as to be entirely valueless.

Connected in no way with manufactures, not having the slightest interest in the existence or non-existence of any particular order of things, the Author trusts he shall supply a desideratum in literature,—and at a juncture, when men's minds are excited upon the subject, which is but just beginning to have that attention paid it, that its magnitude and overwhelming value demand. In the prosecution of his work he has studiously confined himself to data, generally the results of extensive personal research and inquiry, or obtained from documents worthy of implicit belief.

The questions of infant labour, the effects of factory labour in general, the physical condition and peculiar diseases of the labourers, have been treated plainly and succinctly, upon considerations drawn from the known constitution of the human body, and the consequences produced by various external agencies upon its healthy functions. They have been discussed with strict impartiality, and freed from the misrepresentations and gross exaggerations which have hitherto involved them.

The moral, social, and domestic relations of the immense portion of the productive population of the kingdom, now engaged in manufactures, present a picture, as strange and as deeply interesting, as any in the whole circle of the history of mankind. To the philanthropist—to the man anxious for the well-being and social happiness of his species—it is one calculated to fill the mind with sorrow and fear. It has been little attended to, and still less understood. Remedies are wanted—are loudly called for; but, before they can be efficiently applied, the disease must be studied; mere empiricism, even when founded on the purest motives, is dangerous; and when the interest and happiness of a multitude are at stake, a clear and distinct understanding of its wants and failings should precede any attempt to satisfy the one or rectify the other.

In developing the existing state of things, with respect to combinations, the truck system, and the cottage system, many important truths are brought into full daylight, unperverted in their bearings by the mistaken ideas or disingenuousness, sometimes of one party, sometimes of another, which have invariably distorted, or partially concealed them, as best suited their peculiar purposes. A strenuous effort too has been made to demonstrate satisfactorily, how much mischief is arising and has arisen from a division of interest, and want of mutual co-operation between the masters and the men.

In the statistical divisions of this work, the Author has done all that a man can do with the subject and he can add his mite of vexation to that of others who have been engaged in similar inquiries. The utter want of any thing approaching to correct registration, is one of the severest drawbacks to those who are wishful to verify their positions by references to figures. All that the utmost industry can expect to attain, is but an approximation to accuracy, and it may be questioned whether it is not safer to rely upon deductions drawn from other and better authenticated sources of knowledge. The difficulty in these respects is increased a hundred-fold in manufacturing districts, the population of which is undergoing such incessant changes, quite independent of the operation of natural events, and in which too the separatists from the Established Church, constitute the numerical majority. Many of these keep no registries whatever; and in all so much laxity and uncertainty prevail, that they are worse than useless as documentary evidence. The immigration of families from the surrounding parishes is another difficulty, as it generally happens that an individual who has removed not too far from the grave of his fathers, is carried away after death, to repose in the same quiet nook, remote from the turmoil and bustle which have made his heart yearn for a green and sunny resting place.

The little amount of written information on all these subjects, is very extraordinary, especially when it is borne in mind how large a proportion of the national wealth, and of the entire population, are involved in them. This may perhaps have arisen from the circumstance of the existing pre-eminence of manufacturers, being but the growth of a day—being as yet but a Hercules in the cradle. It is indeed only since the introduction of steam as a power, that they have acquired their paramount importance; one generation even now is but passing away since this epoch, and what mighty alterations has it already wrought in the condition of society—changing in many respects the very frame-work of the social confederacy, and opening into view a long vista of rapid transitions, terminating in the subjection of human power, as an agent of labour, to its gigantic and untiring energies.

It is high time that the erroneous opinions currently entertained on many points connected with factory discipline and factory labour should be removed; and that the feelings of sympathy and needless commisseration, expressed by certain portions of the community, should be directed into a more legitimate and less harmful channel. One great cause of the unhappy differences which have so completely sundered the confidence of the employers and the employed, had its origin in the want of diffusion of correct knowledge. Neither have the masters at all times shewn that their opinions were trustworthy. But it has been the men who have chiefly erred, and with the most deplorable results to themselves. They have suffered their better judgments to be warped, their passions led astray, by heartless and designing demagogues,—who have taken advantage of their ignorance, led them to the commission of crimes, have ruined their comfort, and destroyed their social character, for the abominable purpose of battening and luxuriating upon their miseries and degradation.

The causes which have led to the present declension in the social and physical condition of the operatives, the Author believes either to have been wholly misunderstood, or very inadequately appreciated. It is not poverty—for the family of the manufacturing labourer, earn what is amply sufficient to supply all their wants, with the exception of one particular class, the hand-loom weavers: it is not factory labour, considered per se; it is not the lack of education, in the common acceptation of the word:—no; it has arisen from the separation of families, the breaking up of households, the disruption of all those ties which link man's heart to the better portion of his nature,—viz. his instincts and social affections, and which can alone render him a respectable and praiseworthy member of society, both in his domestic relations, and in his capacity of a citizen; and which have finally led him to the abandonment of the pure joys of home, and to seek his pleasures and his excitements in pursuits, fatal alike to health and to moral propriety.

The part of the masters in the great drama, naturally is, and ought to be, the leading one. They are, above all men, interested; and great and almost fearful responsibility rests with them. They have within their power, the capability of doing much either for good or for evil. It is their bounden duty to examine into every thing bearing upon the subject, with the most minute care, with a spirit of strict justice, and with a disposition to ameliorate, as far as is consistent with their own private and indisputable rights. In doing this, they should endeavour so to act, that the confidence of the men may go hand in hand with them, or it will be of no avail. Their name and reputation, like that of Cæsar's wife, should be above suspicion; and till they are at pains thus to elevate themselves—thus to establish their reputation upon the pinnacle of fair fame and honourable consideration, they will be looked upon by the world and by their operatives, as the originators and as the keepers up of a system, which is undoubtedly fraught with many things subversive to the morals, health, and individual independence, of all concerned in it.

The deductions and statements in this work, have been chiefly drawn from and founded upon examinations of that portion of manufacturers, connected with the cotton trade; and this, for several especial reasons. In the first place, this class constitutes of itself about five-ninths of the whole population engaged in the entire manufactures of the kingdom; in the second place, it is gathered together in great and well-defined masses; and in the third place, it may be truly asserted, that "*ex uno disce omnes.*" The universal application of steam power as an agent for producing motion in machinery, has closely assimilated the condition of all branches, both in their moral and physical relations. In all, it destroys domestic labour; in all, it congregates its victims into towns, or densely peopled neighbourhoods; in all, it separates families; and in all it lessens the demand for human strength, reducing man to a mere watcher or feeder of his mighty antagonist, which toils and labours with a pertinacity and unvarying continuance, requiring constant and sedulous attention.

It is much to be wished that "men in high places" would examine, somewhat more in detail, the existing condition of the manufacturers. The conversion of a great people, in little more than the quarter of a century, from agriculturists to manufacturers, is a phenomenon worthy the attention of any statesman. It is an attention, too, which will be forced upon them, when, from their want of knowledge, they will be liable to take steps which may prove destructive to one or both of the great divisions of national property. The change, too, which has recently been made in the constitution of the House of Commons, is another important reason why information should be extended widely and generally. The change is one, indeed, characteristic of the times; and the introduction to the House of persons of strictly commercial habits and views, will necessarily operate very powerfully upon the course of legislation, when brought to bear upon its interests. The change will not, however, rest here: the preponderance already acquired by manufactures, is progressing rapidly; and its representatives will, ere long, take the position afforded them by their vantage ground. The riches of a great and wealthy nation, are pouring into the tide of manufactures, and swelling its ascendency; and it must soon become in appearance, what it is in reality, the directing current of the national prosperity. From the utter absence of information, it is to be feared that crude and badly digested enactments may interfere between the employer and the employed, to the serious injury of both parties, and to the ultimate endangering of the present lofty pre-eminence enjoyed by Great Britain, as the grand focus of manufacturing intelligence,—and may so hamper and obstruct the operations of capital, as to lead, to some extent, to its entire abstraction.

It is, above all things, of importance that the condition of the labourers should be carefully and maturely considered. Though danger may arise from legislative enactments, this is the slumbering volcano, which may at any time shatter the whole fabric to atoms, and involve in one common ruin, themselves, the master, and the manufacture. The struggle carrying on between human power on the one hand, and steam aided by machinery, which is constantly improving in construction, and increasing in applicability, is approaching a crisis, the termination of which it is frightful to contemplate. The contest has hitherto been, between the vast body of hand-loom weavers, who have clung to their ancient habits, till no market is open to their labour; but it is now commencing between another great body and machinery. The adaptation of mechanical contrivances to nearly all the processes which have as yet wanted the delicate tact of the human hand, will soon either do away with the necessity for employing it, or it must be employed at a price that will enable it to compete with mechanism. This cannot be: human power must ever be an expensive power; it cannot be carried beyond a certain point, neither will it permit a depression of payment below what is essential for its existence,—and it is the fixing of this minimum in which lies the difficulty.

It may indeed be questioned how far the interference of Government in questions of this nature is likely to prove beneficial, and whether the parties, the masters on the one hand, and the workmen on the other, would not act most wisely by saying as the merchants of France once said to Colbert, when it was proposed to take measures of protection for their interests—" *Laissez nous faire*," let us alone, we shall take care of ourselves. One thing is becoming daily more and more obvious, and this is, that the agriculturist in the long run has had but little reason to be grateful for the protection which has been thrown over him, not unfrequently at the expense of other interests. It would be well if the manufacturers would take a lesson from them. The majority of representation has hitherto been connected with landed property, and this has enabled the farmer to cover himself with protecting duties, and he has availed himself of it to the utmost. A single glance at his present condition will clearly show how illusory, how fallacious has been his factitious prosperity, and that the time is come when he feels how miserably all his hopes have been blighted. "*Laissez nous faire*" should be the maxim of the manufacturer, or the time will also come when he will bitterly repent his clamour for legislative protection.

London, June 10*th*, 1833.

THE MORAL, SOCIAL, AND PHYSICAL CONDITIONS OF THE MANUFACTURING POPULATION, ANTERIOR TO THE APPLICATION OF STEAM.

Domestic Manufacturers–General union of Manufacture and Farming–Residences and Characters of the Home Labourers—Home Labour, its influence upon the domestic and social virtues—Physical Conditions—Families—Distinction of Rank—Squire, Yeoman, and Labourer—Comparison of the wants and habits of the past and present Manufacturers—Leasing of Land—Peculiar Habits—Sexual Indulgence prior to Marriage—Its influence upon Morals and Character—Manners, &c.—State of Manufactures—Limited supply of Yarn—Inadequacy of existing Machines—Hand-loom Weaver and Farmer—Rate of Wages—Causes which led to the abandonment of Farming as an accessory to Weaving—Influence upon the social rank of the Weaver—Simple hand-loom Weaver, his Condition, his Improvement, and Causes of it—Spinners, their Condition, Changes, improvement in their Machines—Conversion of the small Freeholders into Spinners, their Character—Effects of this upon Farming—New race of Farmers—Establishment of Mills—Influence upon the domestic Spinner—Ruin of the class of Yeomen—New order of Weavers—Human power, difficulty of managing.

Prior to the year 1760, manufactures were in a great measure confined to the demands of the home market. At this period, and down to 1800, during which interval, a general impetus had been given to this branch of trade by foreign and increased home consumption—and in which, also, great improvements had taken place in the construction of the machines, all tending to facilitate and hasten production—the majority of the artizans engaged in them had laboured in their own houses, and in the bosoms of their families. It may be termed the period of domestic manufacture; and the various mechanical contrivances were expressly framed for this purpose. The distaff, the spinning wheel, producing a single thread, and, subsequently the mule and jenny, were to be found forming a part of the complement of household furniture, in almost every house of the districts in which they were carried on, whilst the cottage every where resounded with the clack of the hand-loom

These were, undoubtedly, the golden times of manufactures, considered in reference to the character of the labourers. By all the processes being carried on under a man's own roof, he retained his individual respectability; he was kept apart from associations that might injure his moral worth,—whilst he generally earned wages which were sufficient not only to live comfortably upon, but which enabled him to rent a few acres of land; thus joining in his own person two classes, that are now daily becoming more and more distinct. It cannot, indeed, be denied, that his farming was too often slovenly, and was conducted at times but as a subordinate occupation; and that the land yielded but a small proportion of what, under a better system of culture, it was capable of producing. It nevertheless answered an excellent purpose. Its necessary tendence filled up the vacant hours, when he found it unnecessary to apply himself to his loom or spinning machine. It gave him employment of a healthy nature, and raised him a step in the scale of society above the mere labourer. A garden was likewise an invariable adjunct to the cottage of the hand-loom weaver; and in no part of the kingdom were the floral tribes, fruits, and edible roots, more zealously or more successfully cultivated.

The domestic manufacturers generally resided in the outskirts of the large towns, or at still more remote distances. Themselves cultivators, and of simple habits and few wants, the uses of tea, coffee, and groceries in general, but little known, they rarely left their own homestead. The yarn which they spun, and which was wanted by the weaver, was received or delivered, as the case might be, by agents, who travelled for the wholesale houses; or depôts were established in particular neighbourhoods, to which he could apply at weekly periods. Grey-haired men—fathers of large families, have thus lived through a long life, which has been devoted to spinning or weaving, and have never entered the precincts of a town, till driven, of late years, by the depression in their means of support, they have gone there, for the first time, when forced to migrate with their households, in search of occupation at steam-looms.

Thus, removed from many of those causes which universally operate to the deterioration of the moral character of the labouring man, when brought into large towns—into immediate contact and communion with his fellows, and under the influence of many depressing physical agencies—the small farmer, spinner, or hand-loom weaver, presented as orderly and respectable an appearance as could be wished. It is true that the amount of labour gone through, was but small; that the quantity of cloth or yarn produced was but limited—for he worked by the rule of his strength and convenience. They were, however, sufficient to clothe and feed himself and family decently, and according to their station; to lay by a penny for an evil day, and to enjoy those amusements and bodily recreations then in being. He was a respectable member of society; a good father, a good husband and a good son.

It is not intended to paint an Arcadia—to state that the domestic manufacturer was free from the vices or failings of other men. By no means; but he had the opportunities brought to him for being comfortable and virtuous—with a physical constitution, uninjured by protracted toil in a heated and impure atmosphere, the fumes of the gin-shop, the low debauchery of the beer-house, and the miseries incident to ruined health. On the contrary, he commonly lived to a good round age, worked when necessity demanded, ceased his labour when his wants were supplied, according to his character, and if disposed to spend time or money in drinking, could do so in a house as well conducted and as orderly as his own; for the modern weaver or spinner differs not more widely from the domestic manufacturer, than the publican of the present day, differs from the Boniface of that period, whose reputation depended upon good ale and good hours, and who, in nine cases out of ten, was a freeholder of some consequence in the neighbourhood.

The circumstance of a man's labour being conducted in the midst of his household, exercised a powerful influence upon his social affections, and those of his offspring. It but rarely happened that labour was prematurely imposed upon these;—his own earnings, aided by the domestic economy of his wife, generally sufficing to permit growth and bodily development to be, to some extent, completed, before any demand was made upon their physical energies. This permitted and fostered the establishment of parental authority and domestic discipline. It directed the child's thoughts and attachments to their legitimate objects, and rendered it submissive to that control which is essential to its future welfare. When it was enabled to join its exertions to those of its parent for their mutual support, it did so with no idea of separate interests, but with a free acknowledgment that the amount of its earnings was entirely at the disposal of the head of the family. Thus remaining and labouring in conjunction with, and under the eye of, its parents, till manhood and womanhood were respectively attained, it acquired habits of domestication, exceedingly favourable to its subsequent progress through life; home being, to the poor man, the very temple of fortune, in which he may contrive, if his earnings are not scanty indeed, to live with comfort and independence.

In this respect, the child of the domestic manufacturer was advantageously placed. It remained under its paternal roof during the period in which puberty was developed; its passions and social instincts were properly cultivated, its bodily powers were not too early called into requisition, it had the benefit of green fields, a pure atmosphere, the cheering influences of nature, and its diet was plain and substantial. Under these auspicious circumstances it grew vigorously, acquired a healthy tone of system, the various parts of its physical organization were well compacted and arranged, and it presented an aspect fresh and blooming, speaking of animal energy and vital activity, and evincing that with it every thing went well, and that it enjoyed and profited by them.

The distinctions of rank, which are the safest

guarantees for the performance of the relative duties of all classes, were at this time in full force; and the 'Squire,' as the chief landed proprietor was generally termed, obtained and deserved his importance from his large possessions, low rents, and a simplicity and homeliness of bearing which, when joined to acknowledged family honours, made him loved and reverenced by his tenants and neighbours. He mingled freely with their sports—was the general and undisputed arbitrator in all questions of law and equity—was a considerate and generous landlord—a kind and indulgent master—and looking at him in all his bearings, a worthy and amiable man; tinged, it is true, with some vices, but all so coated over with wide-spreading charity, that the historian willingly drawsthe veil of forgetfulness over them. This race of men is now nearly extinct in the manufacturing districts. Their possessions are passed into new hands—their descendants "pushed from their stools" by an order of men having few or no traits in common with them. They indeed belonged to a race indigenous to a peculiar grade of social advancement, no longer in being. With them, for they were equally a part of the same system, the "Yeomanry of England;" the small farmers have fallen victims to the breaking up of a condition of society, and a state of property, which secured a flourishing rural population. The few old relics of the squirearchy which are still now and then to be found in these districts, are but mementos of a race gone by—are like the scathed oak of another generation, withering in the midst of a thriving and usurping plantation, which has derived its birth and part of its sustenance from its decay, and which is rapidly sinking or already sunk below the level of the mass that surrounds it.

Modern intellectual utilitarians measure men's actions, and every thing referring to his condition, by mere mind. That state of things is, however, the best which brings to each man's fireside the greatest amount of domestic comforts, though it may not be that which will enable him to boil his pot scientifically, or calculate an abstruse problem in mathematics. It is not denied, neither is it wished to conceal the fact from observation, that squire, yeoman, and manufacturer differed very materially in some of these respects from the individuals occupying a correspondent rank at the present period, and that the latter, when looked upon in this point of view, are infinitely superior to the former. Neither in comparing the two is there any intention of wishing that the state of society should be again taken back to that particular level in the onward progress of civilization. It would be vain and childish to do so. The forward impulse must and will be obeyed, whatever the consequences may be to the welfare and happiness of the labouring community. On the one hand may be seen a daily spread of knowledge, joined to a gradual depression in the scale of social enjoyments ; vast and incessant improvements in mechanical contrivances, all tending to over-match and supersede human labour, and which threatens, ere long, to extirpate the very demand for it; a system of toil continued unbroken by rest or relaxation, for fourteen or sixteen hours, in a heated and crowded mill, and an utter destruction of all social and domestic relations; on the other hand, a calm and equable flow of occupation, alternating between the loom and the greensward—an intelligence seldom looking beyond present means—an ignorance of nearly every but the most common arts of life—a knowledge chiefly traditional—a proper station in the social arrangement, a demand for labour full as great, if not greater than the existing supply—a rate of wage quite equivalent to the simplicity and limited range of their wants, and all the social and domestic relations in full force, and properly directed. The great question here is the measure of social and domestic happiness, for these are but synonymes of social virtue. They co-exist and have an intimate dependence one upon another. Degradation in these conditions will ever be simultaneous with moral declension—a declension incompatible with the performance of a great majority of man's best and most sacred duties.

If the comfort of the poor man is to be estimated by variety of wants, by his living in an artificial state of society, surrounded by all the inventions resulting from a high degree of civilization—by having these brought to his door, and every facility afforded him for procuring them—the aboriginal and home manufacturer sinks very low when compared with the present race. If, on the contrary, comfort and domestic happiness are to be judged by the fewness of a man's wants, with the capability of securing the means for their supply, the tables are turned in favour of the domestic manufacturer. If the comfort of these states of society is to be calculated upon another ground—namely, the nature of their separate wants and habits—it brings into light very striking contrasts. The present manufacturer shows a high order of intelligence, seeking its amusement in the newspaper, the club, the political union, the lecture room; looking for its stimulus in the gin and beer shops; for its support a limited supply of animal food, once a day, joined with copious dilution of weak tea, the almost universal concomitant of the spinner and weaver's breakfast and evening meal, in many cases indeed being nearly its sole constituent; debarred from all athletic sports, or rather not having a moment's time to seek, or a bodily vigour capable of undertaking them; an active mind in a stunted and bloodless body; a separation of the labourer from his family during the whole day, and a consequent disruption of all social ties, and this too joined to a similar separation amongst the various members of his household The domestic manufacturer possessed a very limited degree of information; his amuse ments were exclusively sought in bodily exercise, the dance, quoits, cricket, the chace, the numerous seasonal celebrations, &c.; an utter ignorance of printed books, beyond the thumbed Bible and a few theological tracts; seeking his stimulus in home-brewed ale; having for his support animal food occasionally, but living generally upon farm produce, meal or rye bread, eggs, cheese, milk, butter, &c.;* the use of tea quite unknown, or

* The sturdy opposition given to the introduction of tea, by many of the class of yeomen, is strikingly illustrated by an anecdote connected with an ancestor of the author. This man, a respectable landowner and tanner, had, like many others, a wife strongly disposed to have her own way. When the use of tea first became general, she procured a set of the fairy tea-cups and huge silver tea-spoons then in vogue, and got some tea. Many and serious altercations took place between the worthy couple in consequence; the husband steadily refusing what he emphatically denominated stuff, and threatening once and again to demolish the whole paraphernalia if she persisted in bringing it into daylight. However, the good woman and one or two of her neighbours had made the notable discovery that it was a beverage singularly fitted to sit and gossip over. On one particular day the husband had been more than usually furious, and she thought it prudent to beat a partial retreat, but by no means to give up the matter. She was for effecting this by expediting the hour of tea-drinking, so that all should be over before he came to his evening meal. It so happened that on this precise day, she had with her one of her gossips; and during their tea-drinking, by some unlucky chance, Thomas, for such was his name, was heard clattering his thick-soled shoes over the rough pavement before the door, and evidently coming in. All was hurry to conceal her disobedience to his express commands, and this they believed they had effectually compassed by thrusting tray and tea-things upon a stair-head leading to a sort of lumber room, a place which he seldom visited. As fate would have it, he entered and bent

only just beginning to make its appearance; a sluggish mind in an active body; labour carried on under his own roof, or, if exchanged at intervals for farming occupation, this was going on under the eye of, and with the assistance of his family; his children growing up under his immediate inspection and control; no lengthened separation taking place till they married, and became themselves heads of families; engaged in pursuits similar to his own, and in a subordinate capacity; and lastly, the same generation living age after age on the same spot, and under the same thatched roof, which thus became a sort of heir-loom, endeared to its occupier by a long series of happy memories and home delights—being in fact looked on as an old and familiar friend; and in the end, crowding the same narrow tenement in the quiet and sequestered church-yard—there suffered to moulder in peace beneath its fresh and verdant turf, undisturbed by the unholy and sacrilegious hand of the robber of the dead, and swept over by the free, the balmy, and the uncontaminated breath of heaven.

One circumstance exceedingly favourable to the development and continuance of industry and temperance, was the plan, now nearly if not quite

his way to this precise spot, but quite accidentally. The wife expected nothing less than to have to bewail the utter destruction of her crockery. By a rare forbearance he stepped over them, came back, said not a word, and shortly afterwards became fond of it himself. His expression was—"I cock,' lad, I found it no use to oppose it."

exploded, of leasing small farms for one or more lives. A man's character always aided him very materially in these arrangements with his landlord, as upon this depended the amount both of rent and fine, a respectable man, it being well known, not being likely to exhaust and impoverish the land during his interest in it. Numberless examples might be quoted in illustration of this position, and many circumstances detailed showing what a beneficial influence resulted from it, upon the tone and bearing of the manners of the small farmers. They were indeed a highly creditable body of men—priding themselves alike on the good opinion of their landlord, the excellent condition of their farms, and the quality of their yarn or cloth.

The social virtues it has been said will be found *pari passu* with social comforts. The vices that were incident to this state of things were the result of man's natural propensities, and depended more upon the want of any sufficient check upon his passions, which were kept in a state of excitement by health and bodily vigour, than upon moral depravity springing from intentional and reiterated wrong doing. The very fact of these small communities, (for they were generally found congregated in petty irregular villages, containing from ten to forty cottages,) being as it were, one great family, prevented, except in a few extraordinary instances, any systematic course of sinning. This moral check was indeed all-powerful in hindering the commission of crime, aided by a sense of religion very commonly existing amongst them. In one respect this failed however —and it was in preventing the indulgence of sexual appetite, in a way and at a time which are still blots upon the rural population of many districts.

Some surprise may be excited probably by the assertion, in those whose attention has never been directed to the subject, but it is not the less true, that sexual intercourse was almost universal prior to marriage in the agricultural districts. This intercourse must not be confounded with that promiscuous and indecent concourse of the sexes which is so prevalent in towns, and which is ruinous alike to health and to morals. It existed only between parties where a tacit understanding had all the weight of an obligation—and this was, that marriage should be the result. This, in nineteen cases out of twenty, took place sooner or later; but still the prevalence of such a course of indulgence was in some respect decidedly injurious to the character and habits of both individuals; weakening, as it inevitably did, that confidence which ought to exist between unmarried parties, and divesting woman of that chaste and pure influence which is capable of keeping in check man's coarser and more vehement passions. Many of the sports of the period, amongst the young of both sexes, were obviously intended to facilitate and give opportunities to familiarities of the closest kind, and were carried on under the guidance of some old crone whose directions governed the apparent conduct of the females upon these singular occasions. Many curious laws are still in being orally, which throw strong light on the social habits of these children of nature down to a very recent period.

It may be urged that an argument is thus afforded for condemning a system, having as a part of its institutions, or at least if not acknowledged as such, holding out a sufferance for grossness and sensuality alike repudiated by public decency and private morality. The force of this is at once lost when it is known that such a state of things was avowed, and admitted as forming an integral portion of their habits. Morality consists not so much in particular laws, applicable to generals, as upon particular customs, binding particular communities, in which every separate member does that which is done by his neighbour, and in which all share alike the good or evil of them. It would be as unjust to bring forward a charge of licentiousness and immorality founded upon this fact, against the primitive manufacturer and agriculturist, as it would be to condemn as immoral the customs of other nations, many of the habits of half civilized countries, and nearly the whole of those of savage tribes. They do not acknowledge nor understand the conventional laws, which form the code of morals now in being, and are not, therefore, amenable to decisions founded upon them. The labouring population at the period here indicated, did not possess that high tone of public decency, which the advance of refinement has since developed. It must, therefore, be judged leniently in these respects, and not condemned as sinning greatly, because wanting a certain fastidiousness of manners.

Man is nevertheless a rational creature, and the more he governs himself by the dictates of reason, the nearer he undoubtedly approaches to moral perfection. His destiny was never meant to be that of a mere animal, nor was it intended that his happiness should depend upon the gratification of his appetites. Unlimited or unrestrained indulgence of sexual desire in early life, is prejudicial equally to the tone of body and mind, and it is much to be wished, that the custom here alluded to was eradicated and totally forgotten. Bad as its effects were, and are still, they sink into utter insignificance, when compared to the condition of the present manufacturing population in these respects, the consideration of which forms a subsequent portion of this work.

The moralist, though he may condemn and declaim against this, as being at variance with his peculiar ideas, will, if just, attach no farther blame than by saying—it was the error of a system belonging to one stage of civilization. But the philanthropist, the man who, free from morbid sensibilities, views and judges the manners and customs of states in relation only to the happiness of the existing race, will lament—that intellectual excellence should be still attended by similar evils, in an aggravated and more vicious form, hid though they may be, to some extent, by outward refinement—whilst he will detect and deplore the absence of all that warm-heartedness that devotion to home and domestic duties, which in no small degree served as their atonements.

The mischief produced by this means was, however, of small amount. This premature intercourse occurred, as has been before stated, generally between parties, when a tacit though binding understanding existed. Its promiscuousness seldom went further. So binding was this engagement, that the examples of desertion were exceedingly rare– though marriage was generally deferred till pregnancy fully declared itself.*

---

* Poor laws Commission.—" I appeal to the experience of all overseers in rural districts, whether instances of marriage taking place among the labouring classes are not so very rare as to constitute no exception to the general assertion, that 'pregnancy precedes marriage.' "—Extract from Mr. Cowell s report.

It is much to be deplored that the depression in the condition of the agricultural portion of the population should have led to the demoralization which this report exhibits in such decided characters. Unhappily the operation of poor laws and bastard laws holds out a premium for moral depravity. As the report very correctly says—" It would be impossible for the heart of man or demon to devise a more effective instrument for extinguishing every noble feeling in the female heart, for blighting the sweetest domestic affections, and for degrading the males and females of that portion of the community connected with the receipt of parish relief, than this truly diabolical institution." Again, " the English law has abolished

There can be no question, but the more widely inquiries are extended, the more obvious becomes the fact, that the domestic manufacturer, as a moral and social being, was infinitely superior to the manufacturer of a later date. He was much more advantageously placed in all points, and his condition gave ample evidence of this. That he was inferior in some respects is not denied: he could seldom read freely, or write at all; but he went to church or chapel with exemplary punctuality: he produced, comparatively speaking, but little work, but he was well clothed and well fed; he knew nothing of clubs for raving politicians, or combinations which could place him in opposition to his employer, but he was respectful and attentive to his superiors, and fulfilled his contracts to the letter. He had never heard of mechanic institutes, Sunday schools, or tract societies; but he listened devoutly to the reading of his Bible; he had no gin or beer-shop orgies, but he spent his hours in rude sports, in the highest degree conducive to health: he followed, in short, that art mentioned by Cicero: "Hanc amplissimam omnium artium bene vivendi disciplinam, vitâ magis quàm litteris persequuti sunt."

Various circumstances had been for a considerable period producing important modifications in

---

female chastity, self-respect, proper pride, and all the charities of domestic life, connected with its existence. It has destroyed likewise the beneficial influence which this virtue in woman reflects on the character of man."—Vide Report—passim.

the condition of the domestic manufacturer, prior to the introduction of steam as an antogonist to human power; and a brief review of these will form an interesting and curious feature in the history of manufactures, and illustrate the great changes which the last forty years have sufficed to operate in the circumstances and character of those engaged in them. It is singular how little attention has been as yet paid to this subject. A complete revolution has been affected in the distribution of property, the very face of a great country has been re-modelled, various classes of its inhabitants utterly swept away, the habits of all have undergone such vast alterations, that they resemble a people of a different age and generation; and yet little has been said, and still less written upon them; and an epoch which has revolutionized and changed a great nation, has been going on, without any examination of its precursory events, or its ulterior influences.

The distaff and spinning-wheel, producing a single thread, were for a long time the only methods of spinning. About the middle of the last century, the demand for cloth was so much increased, that the inventive faculties of those interested in it were called into activity, and improved machines for spinning were very soon the result. Some of the contrivers of the most important improvements failed to demonstrate for a while, the value of their discoveries; others were driven by a cruel and ill-judged persecution, to carry their knowledge to other countries. A narrow and prejudiced view was taken by the public at large, and little encouragement given for a time, to those who have in the end proved themselves the greatest benefactors to their country. In the teeth of all these impediments, the pressing demand for cloth gradually forced the domestic manufacturer to the adoption of improved modes of spinning; for the difficulty had always lain in producing an adequate supply of materials in a state fit for the loom. One half the weaver's time had generally to be spent in waiting for work. This state of things was however remedied by the mule and the jenny, both of which enabled the spinner to produce a greatly increased quantity of yarn, when compared to that formerly supplied by the wheel.

The hand-loom weaver was of course greatly benefited by all these improvements, without any extra outlay of capital—the loom which sufficed to weave a small amount, being of equal efficiency to turn out a greater. The call for more cloth, and the facilities given in the first processes, removed him from the vaccilating and uncertain conditions in which he was before placed.

A family of four adult persons, with two children as winders, earned at the end of the last and at the commencement of the present century, 4*l.* per week, when working ten hours per day; when work was pressed they could of course earn more—a single weaver having been known to earn upwards of two pounds per week. This however required great industry, and was by no means common; weaving too, when thus exercised, is not easy labour; the position in which the weaver sits is not the best for muscular exertion, having no firm support for his feet, which are alternately raised and depressed in working the treddles. He has thus to depend for a fulcrum, chiefly on the muscles of his back, which are kept in constant and vigorous action—while one order of muscles is employed with little power of variation, in moving the shuttle and beam. These processes when carried on for many successive hours, are very wearying, and the exertion required becomes after a while laborious. The weaver who worked hard therefore, actually toiled—a condition widely different from that of the steam-loom weaver.

One of the first effects of the constant demand upon the labour of the weaver, resulting from a more extensive and permanent supply of yarn, was the gradual abandonment of farming as an accessary, which had been very common with the more respectable portion of the weavers. His labour, when employed on his loom, was more profitable, and more immediate in its return, than when devoted to agricultural pursuits. This necessarily led to the introduction of a new order of farm tenants, men who exclusively devoted themselves to the cultivation of the soil, who looked to its produce as their only mode of acquiring the means of support, and who, in nine cases out of ten, were mere holders at will—a lowering

1731991

in respectabilty, which it is probable will be scarcely understood by those who are not familiar with the feeling and grades of rank in this class or labourers.

This change in the weavers occupation, produced very considerable influence upon him in a social point of view. It decidedly tended to lower his rank and character, by making him a simple labourer, and removing him from that which did then, and which still does confer rural distinction,—the holding or possessing land. This had formerly brought himself and children partially upon the same level with the small freeholder or yeoman, properly so called. It also brought him into contact with this order of men on many occasions,—and remotely, it is true, but still near enough to benefit him, it assimulated his condition with that of the gentry of the neighbourhood: it gave him an honest pride in his capacity, heightened the tone of his general character, and was thus morally beneficial to him.

The great body of hand-loom weavers had at all periods been divided, by a well defined line of demarcation, into two very distinct classes. This division arose from the circumstance of their being landholders, being entirely dependant upon weaving for their support. It has been seen that the weaver, who was also a small farmer, lost something by the increase in the demand for cloth, while, from the same cause, the simple weaver, who had all along depended upon that demand, gained a step in advance, and now found himself on an equality with those who had hitherto been his superiors.

The secondary, or inferior class of artizans, had at all times been great sufferers from the impossibility of supplying themselves with materials for their labour. Considerable vacations were frequently occurring in this respect, and at these periods they underwent very severe privations. This irregularity had produced its usual effects upon their industrial character, rendering them improvident, devoid of fore-thought, and careless in their expenditure. Not being able to calculate, had they been so disposed, upon the certainty that their exertions would be invariably called for, they became indifferent, enjoyed the good whilst it lasted, and starved through the interval as they best might.

It is an indisputable fact, that irregularity in the demand for labour, from whatever cause it may arise, by occasionally throwing the workmen out of employ, and generating idle habits, is one of the most disorganizing and degrading influences which can be brought to bear upon their character. Independently of the diminished rate of wages, and the acrimonious and suspicious feeling which, of late years, have sprung up between the masters and men, the stoppages which have taken place in nearly all the mills in rotation, for periods more or less protracted, sometimes from the fault of one party, sometimes of another, have been full as injurious upon the moral dispositions of the men as the other two put together.

The class of poor weavers were thus instantaneously elevated very considerably in worldly consideration. They were freed from one great cause of depression, which had hitherto prevented all chance of improvement. They now took their stand upon the same ground with the weaver, who had hitherto been a great farmer, and who had come down one degree in the social scale, in consequence of giving up his land for the purpose of devoting himself to the more profitable business of weaving. This amalgamation of the two divisions which had heretofore existed amongst domestic manufacturers, raised on the whole their general character as a body, rendered them more united, and by diffusing one species of labour universally over the whole class, gave them a community of interests and feelings which bound them together.

A very material improvement, therefore, had been gradually operating in both classes of weavers, during the half century immediately preceding the application of steam power. This improvement had not arisen so much from any increase in the rate of payment for labour, as from a market generally understocked, and a constantly increasing production for yarn, which enabled them to work full hours, and of consequently to throw off a regular and sufficient quantity of cloth.

The wages earned when the supply of yarn became regular were amply sufficient to supply all the wants of the weavers and to furnish them with many luxuries and comforts proper to their condition. Even after these had been supplied a surplus remained, which if properly husbanded, would have saved them from the miseries which burst over them when steam-looms were set fairly to work. There seems to be, however, something about the operatives generally unfavorable to the accumulation of money, even when placed under circumstances when it might fairly be supposed that they had an opportunity given them for the purpose. Their savings rarely exceeded a very small sum. It is true, they lived better, in fact, they lived too well, they clothed better, and they did not contemplate the change which was impending over them, and consequently did not "set their houses in order," —but rather looked forwards to a continuance of their prosperity, and pleased themselves with the anticipations of still farther additions to their present comforts. It can excite no wonder therefore, that they, with their limited intelligence— strenuously opposed the steam-looms which they looked upon as usurping what they vainly imagined to be the rights of their industry; nay, pushed the principle still further, and appeared to believe that they had a prescriptive and just claim to a monopoly of this labour.

In the meantime the condition of the spinners,

or as they may well be termed during the prosperous days of hand-manufacture—the aristocracy of manufacturers, was undergoing changes still greater and more rapid than that of the weavers, and passing through a revolution which modified the whole body, and terminated finally in the ruin of the greater portion.

In the early times of manufacturing, the spinner and the weaver were to some extent synonymous—so far indeed, that the processes were carried on under the same roof, and by the same individuals; the distaff, the wheel, and the loom being all called into requisition by a single family.

At a later period, when improvements in machinery progressed rapidly, stimulated as they were by the inability constantly felt to supply the demand for yarn, and as these improved machines at each step became more bulky, more complicated, and consequently more expensive—a division began to shew itself between the weaver and spinner,—the latter throwing aside his wheel and devoting himself to the loom, trusting to external sources for his yarn, from a growing inability to purchase such machinery as would enable him to spin profitably—and the former deserting the loom, and applying his funds to procure the best and most recent contivances for spinning, knowing as he did that a market was always open to his produce.

At this period, that is at the time when spinning was become a separate branch, and when the division between the two bodies was well defined, the spinners were joined by another class of persons, who had heretofore held aloof, from manufacture—and this was the yeoman—the male freeholder now nearly extinct as a part of the social confederacy.

The yeoman had lived generation after generation upon his patrimonial acres—rarely increasing their extent and quite as rarely lessening them. He had, however, failed to keep pace with the onward march of events—had confined himself to cultivating his land precisely in the same way in which it had been cultivated by his forefathers—viewed all innovations as rank heresy, vegetated upon his natal soil, profiting either it or the world but little; but having, notwithstanding many points about him of real value. He was strictly honest in all his dealings—though almost universally improvident, more, however, from want of mental energy and forethought, than from actual extravagance— contented with his lot, and a kind and hearty neighbour—but utterly unable to cope with the crisis which was opening upon him. He had hitherto been surrounded by petty farmers—who had generally eked out their bad management as cultivators by being weavers, and who had served him as bulwarks or breakwaters against the impending storm. These were one by one removed, and their places immediately occupied by a race of men who gave a considerably increased rent, and who by improved modes of husbandry, and by wringing from the soil all it could possibly yield, soon drove the small proprietor from the markets which he had so long supplied. Thus pressed upon by superior industry, and an activity with which he could not compete, the yeoman was driven to embark some portion of his means in the purchase of spinning machines, and before very long, great quantities of yarn were produced by the inmates of old farm-houses, in which previously the most sluggish inertness had prevailed.

Nothing can more clearly shew how very imperfectly this class had used its resources, than the fact, that five-sevenths of those who purchased these machines were obliged to have recourse to a loan, generally a mortgage, to raise the money.

The price of the mere complicated spinning machines was very considerable, and it has been seen removed them out of the reach of the inferior class of weavers. This, aided by other causes already noticed, brought the small freeholder into the field. At the same time monied men began to turn their attention to a branch of trade, the returns from which were very rapid. This brought a farther accession of capital into it, and led to the erection of mills containing a greater or less number of spinning machines, propelled by water power, with the assistance of human labour. These mills were built in places at some distance from the towns, though as near as circumstances would permit for the convenience of markets, and readiness of transport for the raw and manufactured material. Their site in other respects was chosen with regard to two necessary things,—one, the existence of a stream of sufficient volume and permanence,—and the other the neighbourhood of suitable workmen.

These mills were exclusively devoted to the first processes of manufacture, namely, carding and spinning. Their gradual increase very soon influenced the prosperity of the domestic manufacturer—his profits quickly fell, workmen being readily found to superintend the mill labour, at a rate of wages high it is true; but yet comparatively much lower than the estimated value of home labour.

Another cause which tended to injure the private spinner was the incessant and expensive improvements in the construction of machines. Thus the man who one year laid out a considerable sum in the purchase of a jenny of the best and most approved make, found himself, in the course of the year following, so much behindhand, that with his utmost industry he could barely turn out a sufficient quantity of yarn to repay him for his present labour, in consequence of alterations which threw the productive power of his machine into the shade. The price of yarn became of necessity depreciated in proportion to to the quantity produced, which was now more than sufficient to supply the home demand. Finding his efforts thus futile to keep pace with improved machinery, he was compelled either to dispose of his jenny, and this at an immense sacrifice, often indeed at a merely nominal price, or make an arrangement with the maker for an exchange at a rate almost equally ruinous. The number of machines which at this period were thrown back into the market, gave a strong impulse to the growth of the mills; a machine that was not sufficiently perfect for the purpose of domestic manufacture, doing well enough in a mill in conjunction with others, worked at a less rate of wages, and assisted by water power, when its use was valueless to its original owner, who was consequently left in many cases in a worse condition, in a pecuniary point of view, than when he commenced his new vocation, no time having been given him to cover its first cost.

This necessarily led to great difficulties and to embarrassments of no ordinary kind, and was the first step towards the abolition of the small freeholder or yeoman. Their long course of inactivity, and the little diffusion of intelligence amongst them, rendered them incapable of maintaining the struggle with men who had been accustomed to the variations of trade; and whose forethought enabled them to apply remedies, and take such precautionary and anticipative measures as screened them from loss. Their little estates became in the lapse of a few years so encumbered, as to be utterly worthless to them, and a very

rapid and very extensive change took place in landed proprietorship from 1790 to 1810, the period when this transition was in progress.

This declension, though nearly, was not quite universal. The apathy which had so long oppressed and hidden the energies of their character, had failed in quite extinguishing them; and a few of these men, shaking off their slothful habits, both of body and mind, devoted themselves to remedying their condition with a perseverance certain to be successful. Joining to this determination a practical acquaintance with the details of manufactures, personal superintendance, and industry, several of the most eminently successful steam-manufacturers have sprung from this class of people, and have long since become the most opulent of a wealthy community.

There is a singularity attending the rise of some of these individuals, showing very strongly how infinitely superior is personal endeavour to accumulated wealth, when both are devoted to the same purposes. Few of the men who entered the trade rich were successful. They trusted too much to others—too little to themselves; whilst, on the contrary, the men who did establish themselves were raised by their own efforts—commencing in a very humble way, and pushing their advance by a series of unceasing exertions, having a very limited capital to begin with, or even none at all save that of their own labour.*

Vide Chapter I. Early Manufacturers.

The time was now arrived when another great change took place in the relative situations of the two divisions of manufacturers. The improved machines; their increased number; the establishment of mills; the accession of capital—one and all produced a much greater quantity of yarn than could by any possibility be converted into cloth by the hand-loom weaver. A surplus of course remained, which was either sent to the foreign market, or remained as a dead weight upon the maker. The demand for cloth was however still unsupplied, and it became necessary to introduce great numbers of new hands as weavers. So long as the supply of yarn had been limited, or beneath, or just equal to the demand, the weavers had felt but little competition; their wages had consequently remained steady. Now, however, when the outcry for cloth continued, and yarn was abundant, a large body of weavers immigrated into the manufacturing districts—almost the entire mass of agricultural labourers deserted their occupations, and a new race of hand-loom weavers, which had undergone none of the transitions of the primitive manufacturers, were the product of the existing state of things.

This body of men was of a still lower grade in the social scale than the original weavers, had been earning a much less amount of wages, and had been accustomed to be mere labourers. The master spinners, therefore, found them ready to work at an inferior price, and thus discovered an outlet for their extra quantity of yarn. This at once led to a great depreciation in the price of hand-loom labour, and was the beginning of that train of disasters which has finally terminated in reducing those who have clung to it to a state of starvation.

Human power is urged beyond a certain point with great difficulty; and, what is still worse, when great numbers of individuals are in exclusive possession of one particular occupation, the natural caprice incident to mankind, with an occasional inaptitude for regular labour, and a disposition to interfere vexatiously, at times, when its employer is pressed, either by a superabundance, or by a too small demand for his goods, have always rendered it a power difficult to manage, and still more difficult to be depended upon. Nor are these the only unpleasantnesses connected with it: men are apt to acquire notions of exclusive possession, and to hold an opinion that improvements in machinery, which lessen the value of their labour, are wrongs inflicted upon them—are infringements upon their peculiar domain. These evils were keenly felt by the early manufacturers, who had to trust entirely to hand-labour. They were subjected periodically to severe immediate losses, and to the chance of still greater ulterior ones, by the refractory spirit of their hands, who timed their opportunity when the markets were particularly pressing, to urge their claims. They had indeed but little alternative,—and it is quite certain that a crisis was rapidly approaching which would have checked, if not annihilated, the progress of manufactures,—when steam, and its application to machinery,* at once turned the current against the men, and has been since steadily but securely sweeping their opposition to the dust, all their efforts to free themselves from its subjection having been totally unavailing, and worse than useless.

---

* The first steam looms were introduced in 1806, and followed by serious riots and breaking and destroying of machinery.

THE

MANUFACTURING POPULATION,

&c. &c. &c.

CHAPTER I.

CHARACTER AND HABITS OF THE EARLY MANUFACTURERS.

Early successful Manufacturers—Origin of many of these—Their Habits and Character—Homes—Drunkenness—Education: its Effects—Extinction of Sexual Decency—Force of Circumstances—Families of some of the early Manufacturers—Wives—Slowness of social Refinement—Causes of this—Character—Influence of a Mother's Example on her Daughters—State of Society—Sons and Male Connexions—Evils attendant upon their introduction into the Mills—Their licentious Habits and Fate—Pernicious Effect upon the social Condition of the Operative—State of this—Ruin of domestic Virtue—Modes of Prevention.

THE position of an individual who undertakes to relate events cotemporary with himself, is one full of difficulties. He is beset on all sides with errors and prejudices; and it requires not only a sound judgment, and dispassionate examination, to enable him to fulfil his duties to himself and to others in accordance with truth and honesty, but also a capacity for selecting the really important objects connected with his inquiries, and

E

sifting these from the chaff and refuse under which they are often buried. Nor is this all—for it generally happens that his own prejudices and trains of ideas run in the same channel, and are intimately mixed up with those of the particular state or condition of society in which he lives. Few men can isolate or free themselves from the thraldom of the present; still fewer can appreciate correctly, or detail clearly, the origin, progress, and future course of a series of changes—which, although they go on around them, and under their immediate inspection, yet ramify so widely, and extend so far, that they lose the clue of certainty in the mist and obscurity of the distance.

It may be laid down as a maxim, that wherever great bodies of men—whatsoever their rank, and whatsoever the cause which has led to their congregation—are brought together, it leads to an immediate deterioration in the moral condition of some of its portions. Great cities and populous districts have, in all ages, been the foci from whence have emanated, if not great at least numerous crimes.

Man has been variously denominated a two-legged animal without feathers—a laughing animal—an animal endowed with reason—an imitative animal—a building animal—with a number of other definitions, which render it apparent that little accurate knowledge was possessed of his true nature, by those who have thus troubled themselves to class him under one general term.

Whatever else he may be, he is decidedly a gregarious animal; and an universal tendency is shewn by each individual, to identify and attach itself to some one of the divisions into which circumstances have separated and arranged the human race.

Example or custom appears to be one of the most powerful agents in the production of the common actions of life. The various grades of society, from the most elevated to the most debased, are led away by it with as much facility as the mule gang is led by the tinkling of the bell attached to the leading animal, and, in the generality of cases, with as little consciousness as to where or to what it may lead them.*

Moral and social life—the domestic attributes—might reasonably have been believed to be removed from the influence of this agent. They are, nevertheless, those which show most distinctly how indissolubly example and custom are bound to, and lead the ordinary details of existence. One should have supposed that the privacy of home would have made it a little kingdom within itself, having its peculiar laws, and acknowledging none but its natural governors—that it would be a domain shut out from the operation of those influences which determine the external relations of society. But these are, in fact, no safeguard. On the contrary, each home in a community forms

---

* "It is a common vice, not only in the vulgar, but in all men in all conditions, to bend their aim and to shape their thoughts in the fashion wherein they were born."—*Montaigne.*

but a part of the great whole, exhibiting, as in a microcosm, the virtues and vices which are its distinguishing characteristics.

One of the most striking revolutions ever produced in the moral and social condition of a moiety of a great nation, is that which has been consequent to the application of steam to machinery. It is one which will be found to possess many points of great interest to every man who considers the happiness, or looks forward to the probable destiny of that large portion of the productive population, connected with manufactures.

The rapidity with which this revolution has been effected, is not the least remarkable circumstance about it; and strikingly illustrates the truth of the foregoing proposition—that wherever men are congregated in large bodies, their morals must be deteriorated. One great effect of the steam-engine has been, to crowd workmen together; to collect them from parts in which they had hitherto formed portions of a scattered population. But the cause, powerful as it is, has been aided by many others of scarcely less efficacy; and it is these which require elucidation, as they throw a strong light upon the present depraved and debased condition of the class of manufacturing labourers.

Many of the first successful manufacturers were men who had their origin in the rank of mere operatives, or who sprung from the extinct class of yeomen. It has been already explained that this class had been driven, by the pressure of circumstances,* to the adoption of spinning, at the period when trade was undergoing that series of changes which ended in the introduction of steam.

The celerity with which some of these individuals accumulated wealth in the early times of steam spinning and weaving, are proofs—if any such were wanting—that they were men of quick views, great energy of character, and possessing no small share of sagacity; and were by these means able to avail themselves to the utmost of the golden advantages, which were presented to their grasp, from 1790 to 1817, a time when they supplied the whole universe with the products of manufacture.

But they were men of very limited general information—men who saw and knew little of any thing beyond the demand for their twist or cloth, and with the speediest and best modes for their production. They were, however, from their acquired station, men who exercised very considerable influence upon the hordes of workmen who became dependent upon them.

The acquisition of wealth, unfortunately for the interests of all parties, was not attended by a correspondent improvement in their moral and social character; on the contrary, all who had an opportunity of watching its effects, can only deplore and condemn the evil purposes to which, for many years, some portions of it were applied.

---

* *Vide* Introduction, *passim*.

The extreme rapidity with which the returns were made for a considerable period—and this too with an immense profit—and the rich and apparently unbounded prospect that was stretched out before them, fairly turned their heads. In this respect, indeed, they resembled the man in the Arabian Nights' tale, whose eye had been touched with the magic ointment, and which in consequence presented to his mental vision an endless display of wealth. The uproarious enjoyments,—the sensual indulgences, which were witnessed at the orgies of these men, totally unchecked by any intercourse with more polished society, should have had the veil of oblivion drawn over them, were it not that, to some degree, they tend to explain the depravity which in a few years, like a moral plague, spread over the manufacturing population, wherever they were brought together in numerous bodies.

The sprinkling of men of more refined habits amongst the early successful cotton manufacturers, was extremely scanty. Very few who brought large capital into the trade, were fortunate—or even made satisfactory progress. Neither will this fact be considered singular, when it is remembered with whom the battle had to be fought. They had to oppose men who had a practical acquaintance with machinery, and who laboured themselves, assiduously and diligently; whereas the previous pursuits and education of the capitalist, had unfitted him, in some respects, for that rapidity of action and quickness of calculation, which were essentially necessary, if he must keep pace with the daily improvements projected and carried on around him.

Master cotton-spinners and weavers then, at the commencement of this important epoch, were in many instances men sprung from the ranks of the labourers, or from a grade just removed above these—uneducated—of coarse habits—sensual in their enjoyments—partaking of the rude revelry of their dependants—overwhelmed by success—but yet, paradoxical as it may sound, industrious men, and active and far-sighted tradesmen.

Many of these might be found, after a night spent in debauchery and licentiousness, sobered down by an hour or two of rest, and by the ringing of the factory bell, going through the business of the day with untiring activity and unerring rectitude—surrounded too, as they were, by their companions, alike busily engaged under their inspection—again to plunge, at the expiration of the hours of labour, into the same vortex of inebriation and riot.

Meanwhile a great change was progressing in the homes of these men. Wealth brought with it some of its usual accessories. The low and irregularly built farm-house, or the cottage attached to the mill, were exchanged for mansions erected purposely for them—larger, more commodious, and in the most modern order of domestic architecture, furnished in a style of shew and expense—if not of taste—sufficiently indicative of the state of the owner's purse and prospects; and to these were transferred the manners which had unhappily disgraced their late more humble residences.

They did not, it is true, introduce here their low and dissolute companions, in whose society a portion of their time was still spent. The re-unions, however, which did occur, though of a different order, consisting generally of men of their own, or somewhat inferior standing, were fully as debased, and drink—drink was their only amusement and occupation.

Utterly destitute of every thing intellectual, and utterly condemning every thing savouring of refinement, whether in manner or thought, they were in some measure driven to the indulgence of their animal sensations. This was generally sought for in the use of ardent spirits, which roused and maddened them for a time into furious joy, and rendered them unconscious of all that was due to decency or propriety. Thus wallowing in intemperance, little wonder can be excited that other passions were stimulated into active operation; and from their situation, unbounded facilities were offered for their display.

Animal sensations must ever be the predominant occupiers of coarse, uneducated minds. The great advantage of education, when conducted in a way calculated to attain its legitimate end, is—that it makes man less a prey to sense, by giving him other and nobler contemplations, and fixing his regards upon mental exertions and enjoyments, which are diametrically opposed to the grosser demands of his nature. This is the advantage that education possesses when legitimately conducted; but these purposes are not gained by common education, which is too fundamental, and too exclusively intellectual, and which aims apparently at merely expanding the ideas, without troubling itself as to their proper direction. Hence they are left to wander back to their original tendency, aided in their intensity by the very imperfect development given to morals.

The almost entire extinction of sexual decency, which is one of the darkest stains upon the character of the manufacturing population—the laxity in all the moral obligations which ought to exist between the sexes, and the consequent loss of this most important influence in the formation of social manners, may be traced, to some extent, to this period of their history.

The little restraint which existed as to intercourse between the sexes,* before this juncture, may with some degree of justice be pointed out as having laid the ground-work or paved the way for the existing dereliction. This is doubtless correct, in so far that it had broken up the delicacy of feeling, and chastity of deportment, which should distinguish the bearing of the young of

---

* *Vide* Introductory sketch, *passim.*

both sexes. This must be looked upon, therefore, as a prior flaw, and some weight conceded to it; but its influence upon the morality of the parties so erring had not one point of similarity with the grossness which now sprung up on all sides, dating its origin, it is again repeated, in the want of moral discipline on the part of many of the early manufacturers.

Condemning, as every man must, the conduct of these parties, it may be remarked, that the mischief lay in no small degree upon the particular juncture in which they were brought so conspicuously forwards. Their want of education —the animal life they had previously led—the sudden accession of wealth—the contempt in some cases generated for refinement, by the discovery they soon made, that wealth, although burdened with blunt and coarse manners, was still an all-powerful agent for procuring worldly respect—the vanity, which leads men to ascribe results to causes personal to themselves, kept up their original vulgarity, in which they took a strange pride—the facilities for lascivious indulgence afforded them by the number of females brought under their immediate control — the herding together of workmen, the result of the factory system—the separation of man and wife during the hours of labour—the dependence which naturally grew up on the part of the labourers—all these are matters which will serve to explain the immorality which marked the bearing of many, though by no means the whole of these parties.

There were amongst them, indeed, some who possessed a very elevated tone of moral thinking—men whose sagacity clearly foresaw many of the evils which threatened, and which have since fallen upon the operatives from these causes; but the impulse was given—the barriers of domestic virtues broken down—and nothing but a strenuous effort on the part of *all* the masters could have checked and destroyed the evil which was marching about them with giant strides, and with almost irresistible force.*

As these men became better established in their several successes, and as they rose in importance by the extension of their trade, and as, by its continuance and diffusion, other parties, of better education, and more refined manners, joined the manufacturing phalanx, a better order of things gradually developed itself, and worked great and beneficial changes in the characters of the masters.

One of the most permanently injurious consequences resulting from the mode of life led by these individuals, and one which has been severely felt and deplored by them, was its influence upon their families and offspring.

---

* "Custom," says Bacon, "is the chief Magistrate of man's life; men should, therefore, endeavour by all means to obtain good customs."

> "Even handed justice
> Returns th' ingredients of the poisoned chalice
> To our own lips."

Many of them had married before any such brilliant prospect as they now enjoyed had opened upon them; and, of course, had married women of a similar grade with themselves. It is a singular fact that woman, plastic as she is in many of her relations, and readily as she accommodates herself to changes in worldly circumstances, yet, if suffered to pass a certain age, she shews her original coarseness much longer, and much more unpleasantly than her husband. This is abundantly verified by an acquaintance with the families of those manufacturers whose wives have progressed with them through a successful career from a humble origin. The husband will certainly not exhibit the polish of a well-bred gentleman, but his manners will be good, and his information, from a long course of shrewd observation of men and things, will be varied and extensive. He is, in consequence, fitted to move reputably in the rank in which his exertions and industry have placed him. Not so his wife—vulgar in speech, or its vulgarity so badly patched up as to render it still more offensive—coarse in her habits, utterly illiterate, tawdry in her dress, and extravagantly vain of her fine clothes, her fine house, her fine carriage, her fine footman—she is an epitome of every thing that is odious in manners. One redeeming point must be granted her, and that is—she is profusely hospitable; though truth compels the confession, that she is led to this from a sense of vain-glory, not prompted by any nobler feeling or intention.

In some degree this may be accounted for, from the circumstance, that the wife's intercourse with any society more polished than that of her husband and his friends, who use her house merely as a place for carousing and drunken revel, is much later than on his part. His extensive transactions necessarily lead him much abroad, and bring him into contact with a higher order of merchants than himself—men of enlarged intelligence, education, refinement of manners. The conventional forms of society to which he is exposed, spite of his pride of wealth and natural uncouthness, produce their proper influence upon him, and, before long, teach him something of the civil observances of life. Unfortunately, for a long period it ends here. He feels the restraints, to which his good sense teaches him to submit, to be irksome, so that he returns home only to resume his usual vulgar habits, in place of transferring there some of the advantages which he himself has unconsciously acquired. His household, therefore, remains in *statu quo* in all these respects. But the time at length arrives when this is to be thrust into society; and it is then found with all its original coating of coarse vulgarism, inelegance of manners, and entire want of refinement.

It was not till a very recent period that any thing approaching to decent society was established amongst these wealthy individuals—a circumstance chiefly arising from the quality of the females, who were, in fact, as little at home in their costly drawing-rooms and boudoirs, as Omai when first clothed in an English dress. They were, indeed, much more frequently to be found in the kitchen than in their present appropriate apartments.

The influence of a mother's habits upon the character and disposition of her daughters, is of so durable and decided a nature, that no wonder can be felt, that many of them—highly cultivated as they have been—educated in the most expensive schools, and separated, for a time, from home—still show traces of the germ from which they have sprung; and that the tone of society prevailing amongst them, differs widely from that—in which mental and bodily cultivation has proceeded hand in hand with the acquisition of wealth; and still more remotely from that of the aristocracy and gentry, upon whose heels, in point of worldly consideration, they are so closely treading.

It was, however, upon the sons, younger brothers, and immediate male connexions, that the example set by these individuals exerted the most pernicious influence. Reversing that line of Horace which says

"Instillata patris virtus tibi,"

it describes very accurately its effects. The demand for hands necessarily led to the employment of all the male relatives of the master manufacturers, at a very early age. His own sons were invested with considerable authority, when mere children—taken from school to superintend certain portions of the mills, and liberally supplied with money. The same remark holds good with reference to others, his relations. Boys, at an age when they should have been sedulously kept apart from opportunities of indulging their nascent sexual propensities, were thrust into a very hot-bed of lust, and exposed to vicious example, in addition to other causes, irresistibly tending to make them a prey to licentiousness. The consequences of this criminal and unadvised proceedings were, that long after the masters had freed themselves from the vices incident to their first advancement, they had the shame and mortification of seeing their errors propagated through a series of ramifications, every remove seeming to become more and more depraved in its character.*

The organised system of immorality which was pursued by these younger men and boys, was

* "The fact then undoubtedly is, that the licentiousness which prevails among the dense population of manufacturing towns, is carried to a degree which it is appalling to contemplate, which baffles all statistical inquiries, and which can be learned from the testimony of personal observers."—*Inquiry*, &c., p. 25.

extremely fatal to the best interests of the labouring community. Chastity became a laughing-stock and byeword. Victim after victim was successively taken from the mill—the selection being, in the generality of cases, the prettiest and most modest-looking girl to be found among the hundreds assembled there. One after another they yielded, if yielding it deserves to be called.

So depraved, however, was their idea of woman's honour already become, that an improper intimacy was rather esteemed creditable than otherwise. The miserable creature was pointed out by her companions, as being peculiarly fortunate in having attracted the notice of the young master, his nephew, brother, or cousin, as the case might be ; while she, in her turn, displayed, with ostentatious parade, the ribbons, cap, or gown, (in reality the insignia of her infamy,) which were considered by her as a rich reward for whatever favours she could bestow.

> " O miseri quorum gaudia crimen habent."

Houses were established in some localities by parties of young men purposely for the prosecution of their illicit pleasures, and to which their victims repaired—nothing loth, it is true—to share the disgraceful orgies of their paramours ; and in which scenes were enacted that even put to the blush the lascivious Saturnalia of the Romans, the rites of the Pagoda girls of India, and the Harem life of the most voluptuous Ottoman.

Numbers of these infatuated youths reaped their reward early, and were cut off on the very threshold of manhood, by a career of unbridled indulgence; others had a more prolonged existence, a "sort of life in death," and sunk gradually into the grave, borne down by slow and lingering consumption; or their physical energies were so reduced, their healthy stamina so destroyed, that they had no power to resist disease, however slight, and were consequently carried away in rapid succession—

"Et minimæ vires frangere quassa valent."

Marriage was repudiated amongst them, and was indeed utterly at variance with the license reigning over all their sexual appetites.

Nothing can more clearly shew the demoralizing effects of this pernicious intercourse than the fact, that a girl, who was known to have lived in a state of concubinage, found no difficulty in marrying subsequently amongst her equals. So debased became their tone of thinking, and so utterly were they lost to all sense of decency and shame, that not only was no difficulty found by these girls in procuring husbands, but this was even managed during the very time their intimacy was going on with their seducer, and which, not unfrequently, was continued up to the very day of their marriage, and even subsequently.

Who shall declare that it was confined to one object, or that the profligate husband, who permitted or sanctioned such abominable proceedings, did not shut his eyes to all comers, going even one step beyond the Roman Gippius, whose remark became proverbial in his own day—"*non omnibus dormio.*"

This laxity, or rather this entire absence of all regard for moral obligations relating to sex, extending, as it did, beyond the threshold of marriage, sapped the foundations and overthrew the structure of the social virtues. Father, mother, son and daughter, were mere words, leading to none of those delightful associations which are the fount from whence spring, in a tide of sparkling purity, the best and most hallowed of man's pleasures and emotions. The links of domestic affection—the passions which properly belong to it, are

> "As gentle
> As zephyrs blowing beneath the violet,
> Not stirring its sweet head;"

and the loss of these has shorn the homes of the manufacturing population of almost every thing which can give home a value.

It must be understood that in pointing out a state of things, and a mode of life, happily greatly amended, the first manufacturers of this epoch must bear the blame for presenting an example, to which the peculiar circumstances of the times gave great and immediate force. Had a high sense of moral obligations taught them to introduce strict discipline into their mills, by separating the sexes as far as possible, and by keeping a watchful eye upon their own passions, a very different aspect would now be undoubtedly presented by this population. The impulse being given, the barrier of private decency being broken down, man's natural propensities, aided in their operation by collateral causes, will long prevent a filling up of the breach; and till this is done, till the social virtues, till the morals of home are established among them, no nostrums of political economy, no bare intellectual education, no extension of political rights, ever can or ever will make them a happy, respectable, or contented race of men.

Many other circumstances, of almost equal potency, tending to demoralize this portion of the population, were brought into operation by the steam-loom and by steam-spinning.

CHAPTER II.

INFLUENCE OF TEMPERATURE AND MANNERS UPON PHYSICAL DEVELOPMENT, &c., AND UPON MORALS.

Puberty and sexual Appetence—Causes influencing their development—Vicious Example—Peculiarity of sexual Passions—Importance of propriety of Manners—Want of Attention on the Part of the Mill-Owners and Overlookers—A Want of proper mental Growth—Error of Mothers—Influence of Temperature on physical and sexual development—Influence of Manners upon physical and sexual Development—Late Acknowledgement of Puberty in Great Britain—Its influence upon physical Condition, &c. and upon Society—Importance of proper sexual Restraint—Value of Refinement of Manners—Temperature, &c.—Necessary Measures on the Part of the Masters—Present sexual Character in manufacturing Districts.

THE crowding together numbers of the young of both sexes in factories, is a prolific source of moral delinquency The stimulus of a heated atmosphere,* the contact of opposite sexes, the example of lasciviousness upon the animal passions—all have conspired to produce a very early development of sexual appetencies. Indeed, in this respect, the female population engaged in manu-

* The average temperature of the atmosphere in mills is about 70 to 75 Fahrenheit; formerly it was much higher.

factures, approximates very closely to that found in tropical climates; puberty, or at least sexual propensities, being attained almost coeval with girlhood.

The influence of these maturing causes is more strikingly seen in the female than in the male, in consequence of its effects being of a more prominent and observable character.

It may be questioned whether puberty, in the common acceptation of the term, is necessary for the display and exercise of functions of sex.* Ob-

---

* "The age of puberty amongst the Tahitians is about ten or eleven. The women have such early connexion with the other sex, that the catamenia never (that I heard of) appear before such connexion.

"Some marry as young as ten; but the average age is about fourteen or fifteen. Marriage takes place from twelve to sixteen years; but formerly sexual intercourse began much earlier, and was often practised when the parties could not be more than six or seven. I am quite confident I am stating fact on this point."—*Answers of Mr. Ellis and Mr. Bourne. Roberton, North of England Medical and Surgical Journal,* No. I., p. 180.

"In Asiatic manners there is much to be found that is odious and feculent; and of this the most pre-eminently disgusting part is the precocious acquaintance with the sexual passion which almost universally obtains. In allusion to the way in which the Hindoos train their children, the Abbé Dubois informs us that, from their earliest infancy, they are accustomed to scenes of impropriety—that it is not uncommon to see children of five or six years of age become familiar with discourse and action which would make modesty turn aside. The nudity in which they live for the first seven or eight years—the loose conversation, tales and songs to which they have to listen—the stories of the dissolute lives of their gods—the solemn festivals so often celebrated, where modesty is totally excluded—the abominable allusions which many of their daily practices recal—

servation has taught, that the usual marks accompanying and preceding this epoch in the female, have very frequently been wanting in instances where unequivocal proofs were given that the individuals had not only suffered, but also enjoyed intercourse. Not only this, but in some examples pregnancy has occurred, where the usual physiological data, generally esteemed requisite for its production, have been entirely wanting.

Independently of this however, the usual accompaniments of puberty shew themselves very early, stimulated into premature activity by the above causes, and unchecked in their effects upon manners by the absence of moral discipline, whether at home or abroad.

So far indeed is this from being the case, that at the mill the factory girl is subjected to a series of excitements, which inevitably lead to the ultimatum of desire. She can hardly be said to have a home—certainly she has none in the proper sense of the word—for the whole of the day is passed from it, and she visits it merely for the purpose of rest. It has no moral ties upon her; and her parents are perhaps positively injurious to her in this respect.*

---

their private and public monuments, on which nothing is ever represented but the most revolting obscenities—and their worship, in which prostitutes act the principal part:—these are some of the polluted sources from which the youthful imagination of the Hindoo draws its imagery."—*Ibidem,* p. 186.

* "After all, what motive has either sex, in the class and

The advantages resulting from the simple observances of common propriety, may be judged of by the evils which attend upon their absence. Gross language, unrestricted allowance of word and deed pertaining to sexual gratification, are pregnant proofs of what the reality is, when the outward form is so debased. The peculiar qualities of these passions—one of the most striking of these being the readiness with which they are roused by a word, a look, or a gesture, and the undying hold they take on the imagination, when once permitted to revel at large—render the want of some sufficient visible check very obvious.

Unfortunately no regard, however slight, is paid to these matters by the majority of mill-owners. A certain number of hands are required to superintend the labours of their untiring engine, with its complement of looms, &c. &c., for a cer-

---

situation to which we allude, for being virtuous and chaste? Where they are unshackled from religious principle, as is too generally the case, they have no delicate sentiments of morality and taste to restrain them from gratifying every passion: they have few or no pleasures beyond those which arise from sensual indulgence—it involves no loss of character, for their companions are as reckless as themselves—*it brings no risk of losing their employment, for their employers know that it would be unsafe to inquire into these matters:* it is often a cause of no pecuniary loss, for in many cases the poor laws provide against this; and all these circumstances considered, the licentiousness of the manufacturing population is a source of bitter lamentation to us, but of no astonishment whatever."—*Inquiry, &c. into the Condition of the Manufacturers, &c.* p. 26.

tain number of hours, and for a certain amount of wages: so long as these are attained, he looks no farther. He considers the human beings who crowd his mill, from five o'clock in the morning to seven o'clock in the evening, but as so many accessaries to his machinery, destined to produce a certain and well-known quantity of work, at the lowest possible outlay of capital. To him their passions, habits, or crimes, are as little interesting, as if they bore no relation to the errors of a system, of which he was a member and supporter.*

Added to this, those who hold the higher rank in his mill, from the overlooker downwards, are all tinged with the same carelessness or coarseness—and in place of being checks and stays, rather lead the train of mischief. Springing from the diseased body themselves, they too often show their origin by a course of conduct to those over whom they are placed, calculated for the gratification of their satyr-like feelings, and the baseness of their own minds.

---

* "If the master acknowledges no common bond as existing between him and his labourers; if he does not even know their names or faces; if avowedly or practically, which is the same thing, he disclaims all regard to their conduct except as manufacturers; if, in fine, he keeps wholly aloof from them—and under present circumstances it is not easy to see how he can act otherwise—then it is clear that some of the best feelings of our nature can never be called into exercise in the breast of the operative."—*Roberton on the Health of Manufacturers*, p. 23.

"Their employers know it would be unsafe to inquire into these matters."—*Inquiry, &c.* p. 26.

The early age at which sexual development calls into play a crowd of irrepressible sensations—which, when properly tempered and directed, form the basis of future character—and the unfavourable circumstances under which this forced development occurs, are, in a great measure, destructive to the well-being of those, who may well be called its victims.

The mind does not keep equal pace with the body, though in some respects its faculties are precocious. Its better qualities are destroyed by the preponderance of animal sensations: it is, at a period when, even under the most favourable auspices, it is vacillating and uncertain in its determinations—now dreaming of sensual indulgence, and now devoting itself to the better purposes of which it is capable—hurried away by external associations—forming itself on the model of whatever is near it, and taking its impress—an impress which will never be entirely worn off from the hands of those, who may accidentally have the fashioning of it.

> "Udum et molle lutum est, nunc properandus et acri,
> Fingendus sine fine rotâ."—*Persius.*

The evil unfortunately does not end with the party first yielding to temptation: were it an isolated case, it might do so; but here, extending as it does, throughout a whole people, it descends from parent to child as an hereditary curse.

The mother who has never felt her own moral

and social rank injured by her sexual indulgences—who looks around her, and sees that all are like herself—who has experienced no difficulty in settling herself as a wife—who even, if, after her marriage, she has continued her former practices, has derived positive and substantial benefits in consequence, by improving the condition of her husband, and adding to the comforts of her family, forgets—if she ever felt—that she was sinning.

Her family inherit the same lax feelings; her sons and daughters are both subjected to the same causes which prematurely evolved her own propensities; are themselves in the same state of precociousness; have the same failings, and become fathers and mothers in their turn.

The additions which are incessantly made to the population of the manufacturing towns, and which additions are made up chiefly of increments from the surrounding agricultural districts, are very rapidly assimilated, as to moral perversity, with those amongst whom they become sojourners.

It is true that those who have already attained maturity, cannot undergo the same physical changes which have determined the moral condition of the parties whose early youth has been spent in the enervating atmosphere of a large mill: but their union with those who have suffered these changes, soon obliterates any difference which might originally have existed between them: or if it should fail in doing this, as far as regards the senior members of these immigrations, it never fails to do so with the juniors. In the end, therefore, although each addition modifies, to a limited extent, the universality of the evils complained of,—these, constant as they are, produce no permanent impression upon the leading traits of the whole community. The rapidity with which generation succeeds generation, and the continuance of the same energetic and efficient causes, operating upon the moral framework of society, must ever prevent any decided improvement—providing this improvement is left to the natural course of events, unaided and unsustained by remedial measures of a more decided character and more cogent operation.

The influence of temperature upon physical development has been tolerably investigated, or at least fully acknowledged, by those who have paid attention to external causes operating upon the human race. Travellers in all tropical countries have rendered the early age at which sexual intercourse takes place sufficiently familiar; and this has been generally taken as a sort of acknowledgment of an equally early puberty. That this is influenced to some extent by climate cannot be doubted; but it may be seriously questioned whether this has not been greatly overrated.*

* If temperature were the sole cause in producing early sexual development, it might reasonably be concluded that the inhabitants of the extreme north would exhibit a very pro-

If the influence of temperature has been exaggerated upon these points, there is one question which has been neglected, in taking into account the causes which govern sexual maturity; and that is, the customs, habits, and modes of life which prevail in the countries in which observations have been made. These are, unquestionably, far more important agents than mere temperature, both in calling into being sexual desire, and accelerating, in consequence, the period of puberty. It is quite needless to enter into any argument to prove — that whatever excites the generative organs, will have a tendency to develope their specific functions; viz. menstruation and conception. The same remark is applicable not only to these organs, but to any and to every organ composing the human body. An attentive examination of the writings of various travellers and

---

tracted virginity. That this is not the case is abundantly proved by northern voyagers and travellers—witness Hearne, Franklin, Parry, Lyon, Crantz, Richardson, &c. &c. Parry declares that in no country is prostitution carried to a greater extent than among the Esquimaux—that no people are more libidinous and dissolute. Mention is more than once made of girls of ten years of age being taken as wives. Mr. Richardson remarks, " that the women marry very young." Examples are given of wives of fourteen, sixteen, and eighteen years of age. —*Vide Parry's* first and second voyages; *Franklin's* first and second journey; *Hearne,* &c. &c.—Humboldt, in speaking of the tribes of northern Asia, says, " that girls of ten years are found mothers." The evidence of Clarke and other writers who have visited the northern countries all agree in the same facts.

observers,* would demonstrate the accuracy of this opinion very sufficiently, and would shew that sexual indulgence precedes and leads to what is here considered premature physical perfection. This is dependent, not so much upon any constitutional peculiarity, as upon manners which encourage and keep up libidinous feelings—and examples brought to bear upon passions so excitable and active as those having reference to sexual enjoyments are.

It has long been noticed, that in warm countries, and in nations still barbarous, or only in a state of partial civilization, very early marriages take place, and that women become mothers at a time which appears singularly premature to northern ideas; at a time, indeed, when girls are looked upon as mere children, alike unknowing of and incapable of being roused to sexual appetence. The moral and religious institutions which have prevailed in Great Britain, have discountenanced the intercourse of the sexes, till, comparatively speaking, a late period of life. Jurists and legislators have also lent the weight of their authority to the protraction of adult age till such

---

* *Vide* Russel's "Natural History of Aleppo;" Marsden's "Sumatra;" Winterbottom's "Natural History of Sierra Leone;" Ellis's "Polynesian Researches;" Buchannan's "Journey through Mysore, &c." Crawford's "Indian Archipelago;" Long's "Jamaica;" Collins's "New South Wales;" Forster's "Voyage Round the World;" Cook's "Second Voyage;" Raffles's "History of Java, &c."; Volney, Madden, and a host of others.

time as it was supposed both body and mind would be equally matured.* These arrangements are, however, merely conventional; but they have nevertheless produced very decided effects upon the physical nature of the population, by acting as so many powerful checks upon the display and early development of puberty. The germs of its maturity were nevertheless in being; and, were these checks removed, it is unquestionable that a great and striking change would take place in the social arrangements at present existing.†

One of the most important advantages which has attended the past and present delay in acknowledging and countenancing the occurrence of puberty till so late a period, has been, that time and opportunity are given for the growth of those feelings of reserve and delicacy which both sexes have been necessarily driven to adopt for

---

* The English law makes it felony to have intercourse with a child under ten years of age; and a misdemeanor if above that age and under twelve.

† The experience of medical men engaged in manufacturing towns affords evidence of this, for in these situations the restraints of law and decency are unacknowledged. A case is related in Mr. Roberton's paper in the *North of England Medical and Surgical Journal*, as having occurred to a respectable surgeon, where a girl working in a cotton factory had become pregnant during her eleventh year, and that in all respects she was a full grown woman, and that the menses had appeared prior to this. Two similar cases have occurred to the author, and very many others to his professional friends. It is far from uncommon to find mothers at fifteen years of age, a period perhaps full as early as that in tropical countries.

their mutual protection. It is this protection dependent upon, and having its origin in, the causes just mentioned, that eventually becomes the armour shielding the chastity of sex: it is this, too, which forms the grand distinguishing mark between a nation far advanced in civilization and refinement, and one commencing its career, or still wallowing in the coarseness of savage existence: it is this which gives its charm to the associations of common life, shutting out and completely excluding the grosser topics, which are subversive to so many of the social virtues—for virtues they are, even though matters of form. By thus ridding the communion of all grossness, the best guarantee is given for its moral propriety. The character of the sexes is raised in the estimation of each other. It clothes them with a covering of purity, which must be first removed, before man's animal nature can be fully displayed; and this is generally prevented by the high tone of public and private morality. Or if, in spite of these, it is rashly cast aside, the wanderer is condemned to the persecution of a host of troubles and vexations, which in the one sex ends by its expulsion from its natural rank in the social confederacy.

In speaking thus strenuously on the observances of the forms of society, it may perhaps be imagined that too much stress is laid upon them, and that the natural instincts of mankind, its moral tendencies, with the still higher principle, the "mens divinior" are too little consulted—in fact, that the composition of the essence is overlooked in its fragrance and beauty.

The form here, however, constitutes whatever there is of reality. Man cannot be taught to forget that he is a man, or that the breathing and blushing being before him is a woman; that she is endowed like himself with an ardent temperament—a desire for gratification; that she has within and around her a world of delights which he is framed for and destined to enjoy; and that she has passions which, if roused into activity, would overwhelm all sense of shame or propriety. Neither can he forget that he has a fire glowing within his own breast, that, if freed from the asbestos coating of moral decency, would overthrow all obstacles standing between him and the object of his desire,—nor that he has the capability of stirring into vigorous life, his own and woman's propensities. These he cannot be taught to forget—the voice of nature is potential. But he may be taught that his own welfare, and more especially that of the creature before him, are intimately connected with the suppression and concealment of their separate desires and capabilities. His sense of justice, as to what is due to himself, to her, and to the still wider circle in which he moves amongst his fellows, may be called into action, and thus become his own safe-guard, and the safe-guard of the woman, against the machinations of their respective passions; and hence it is that the forms of society are worthy the greatest attention, and the most sedulous cultivation.

It is impossible to adduce any fact more clearly elucidating and corroborative of these remarks, than the state of society in Paris now and for many years past. Her literati and doctrine-mongers, from Voltaire, their apostle, downwards, have gone on step by step, to strip the relations of life of all the drapery of true delicacy. One by one, its delightful and soothing illusions have been exposed to the batteries of sarcasm or held up to ridicule as hiding man's natural character, and thus making him an artificial rather than a real being. This purblind philosophy has succeeded in producing a social condition infinitely inferior to that of the savage,—the naked and unadorned morals of which it has tricked out in the tinsel glitter of intellectual excellence—only to increase and render more conspicuous its deformity.

A similar state in the social and domestic relations has been brought about in the manufacturing population, more especially in the large towns, but by very dissimilar causes. Here it is the triumph of sense over morals—a great revolution in the period of physical development,—a consequent premature indulgence of sexual appetite, unchecked by any sufficient impediment, and producing important effects upon all that relates to the morals and well-being of those subjected to their influence.

The change in the moral and physical attributes of this population have been operated by temperature partly, but more especially by the force of example and manners already existing. Both are no doubt powerful agents, but there can be no difficulty in saying which must bear the principal onus of imputation. Co-operation has manifestly aided the effect of each separate influence, and has been the reason of its unexampled and rapid spread, and its permanence must of course depend upon the continuance or abstraction of the same causes.

Temperature has done something to those who had been exposed to it from infancy, for from this period the factory labourer dates his labour. The manners of those around him have done much more in determining the precocious development of sexual desire and of puberty. The mere fact of having been exposed from the earliest dawn of perception to sights and sounds stimulant upon passion, will easily account for this. Upon those who are introduced to the mill at or posterior to the usual age of puberty, temperature cannot do much; but example, by producing a rush of new sensations, will be all powerful. In the one case then temperature aided by manners has been the cause of the change which has been produced; in the other, manners and example have been the agents for perpetuating and extending it, and reducing all that come within their influence to the same condition. Most unfortunately as yet, no moral or religious check or impediment is in force to prevent the displays, necessarily resulting from this change, —nor are there any modifying and correcting agencies to lessen its mischievous tendencies.

Had it been the wish of the early manufacturers to have made their workmen an orderly, moral, and domestic race, their own example would, in the first place, have been of the utmost importance. Failing in this, the next most important point would have been to have entrusted the overlooking department to individuals of good character and untainted morals, and of humane and considerate dispositions. Then the separating, as far as possible, the sexes, during the hours of labour, and carefully suppressing any display of grossness and immorality, so long as they were confined to the mills. Great improvements have and are taking place in these respects. Next, dismissing or disgracing any party guilty of improper pratices; and finally, rejecting steadily, and as an inviolable rule, all and every applicant for work who did not bring some testimony that he was a sober and moral person.

These measures would have done much towards counteracting whatever forcing effects the temperature of their mills might have. They would, at all events, have prevented them from becoming obnoxious. They would, by taking away other and more influential causes, urging on early sexual and physical development, have partially removed the evils ever attendant upon the crowding together great bodies of people. Powerful restrictions would have also been placed upon that relaxation of sexual decency—that abolition of delicate and refined sentiments, and social obligations, which ever lead to barbarous modes of living, and to the ruin of many of the fairest portions of man's character.

They have, however, been neglected or overlooked, and a power which the masters might have exercised easily in the first instance, it would now be difficult if not impossible to assume. The numerous causes of dissension and suspicion which have arisen between them and the labourers, are so many impediments in the way of beneficial and salutary inspection; till these are removed, little expectation can be had of any striking amelioration, proceeding from their interference. There are, however, certain changes in progress, that will, ere long, most probably enable them to do this easily and efficiently if, in the mean time, the volcanic elements composing the manufacturing population, do not burst out into active operation, and destroy the whole system of which they form the foundation.

As it is, the baleful influence of the circumstances already detailed, are in full and undisguised action; general licentiousness and illicit intercourse* shewing themselves at a very early age, and to an extent painful to contemplate. The exhibitions of the utter absence of sexual delicacy, common

---

* "On the subject of the general licentiousness and illicit intercourse between the sexes, which prevails in manufacturing districts, we cannot, for obvious reasons, dwell so long, nor as minutely, as the extreme importance of the subject would justify. In the few words we shall devote to this branch of our investigation, we shall be careful to keep within the limits of the most scrupulous accuracy, and to affirm nothing which we do not possess the materials for proving. First, then, we shall remark, that nothing but personal observation, or the testimony of eye-witnesses, can be relied on for satisfactory information. The returns of illegitimate children are worse than useless—for it will be obvious, on a few moments' consideration, that in such cases they can afford us no possible criterion of the desired result. On this subject, some writers on political economy† betray the same ignorance as in the assertion of the extensive use of animal food among the manufacturing labourers. Both instances furnish an illustration of what appears to be a common source of error with them—namely, a disposition to draw inferences from isolated facts, instead of resting their doctrines upon the basis of extension and accurate observation. They conclude, that because the proportion of illegitimate births appears to be greater among agricultural than among the manufacturing population, the females of the former are the more immoral of the two. We draw, without doubt or hesitation, exactly the opposite conclusion, and every one intimately acquainted with the south of Lancashire will bear us out in this opinion. The deduction we draw is also materially confirmed by the practice which, it is painful to state, is far from uncommon among the abandoned women of these districts, of destroying, prematurely, the fruit and evidence of their guilt."—*Enquiry*, &c. p. 25.

† Edinburgh Review, xci.

in manufacturing towns, are notorious to the most superficial observer. The robe of Arthur's mistress would be tried in vain on the great majority of the females.

It is not denied that there are many girls who, from a coincidence of favourable causes, or from their possessing a higher and more just sense of what is due to themselves, escape some of the many evils which beset them, and grow up decent and moral women, fitted to make good wives and good mothers. These are, nevertheless, exceptions; not so rare as to be extremely remarkable, but still standing out in strong relief from the mass to which they are attached.

The causes which are so injurious to the female, are operating, in a similar degree, and with similar force, upon the male, and produce precisely similar effects, both upon his moral and physical organization.

CHAPTER III.

THE INFLUENCE OF THE SEPARATION OF FAMILIES, &c. UPON THE MORAL AND SOCIAL CONDITION OF THE MANUFACTURING POPULATION.

Domestic Manufacturer—Union of his Family—Its Consequences and Advantages—No Child-labour—Effects of this upon the Male and Female Portion of his Household described—Factory Labour—Destruction of Home—Separation of Families—Effects of these upon social Character—Payment of Wages to Children, its Evils described—Conversion of Homes into Lodging-houses—Loss of domestic Virtues, description of—Early Period at which Factory Labour begins—The Mind, &c. of Childhood—Division of Families—Cultivation of Home Affections—Infancy of the Factory Child described—Comparison of the Condition of the Agricultural Labourer—Its Advantages—Domestic Economy—Non-existence of, in the Factory Woman—Separation of Man and Wife.

It has been truly remarked by Bacon, that "the culture and manurance of mind in youth, hath such a forcible though unseen operation, as hardly any length of time or contention of labour, can countervail its influence."

The domestic manufacturer possessed one great advantage over the factory labourer, which was —that his occupation was carried on beneath the roof of his own cottage, and in the midst of his family: also, that his children, growing up under his own eye, and around his fire-side, retained for him the respect and awe due to parental authority, by remaining members of one home and under the direction of one head.* By keeping up this natural and proper order of things, he secured one means of making his offspring domestic in their habits, and it was his own fault if their social character was not what it ought to be.

So long as families were thus bound together by the strong link of interest and affection, each member in its turn, as it attained an age fitted for the loom, joined its labour to the general stock, its earnings forming part of a fund, the whole of which was placed at the disposal of the father or mother, as the case might be; and each individual looked to him or to her for the adequate supply of its wants. No separate or distinct interests was ever acknowledged or dreamt of. If any one, by superior industry or skill, earned more in proportion than another, no separate claim was made for such excess on the part of that individual; on the contrary, it was looked upon equally as a part of the wages of the family,—perhaps gratefully and affectionately acknowledged, but leading to no other result.

This family compact, of course, existed no lon-

---

* Vide Introductory sketch—passim.

ger than the usual period when parental control yields before the maturity of offspring. This was rarely before twenty-two or twenty-three years of age, and often much later. Grown up, as each member had, as part and parcel of a little community, these divisions seldom took place before marriage opened a series of new cares and new prospects to son or daughter, which, in consequence seceded, or, as was frequently the case, brought a wife or husband to be joined to the family union. Generally, however, at this period an offset or branching took place, which was best for all parties.

This preserved in all their vigour the moral obligations of father and mother, brother and sister, son and daughter, and that till a time of life was gained, which had given abundant opportunity for the formation of character—a character most assuredly the best calculated to render the labouring man happy and virtuous, viz., a domestic one; without which, no adventitious aid can ever secure him their possession.

The greatest misfortune—the most unfavourable change which has resulted from factory labour, is the breaking up of these family ties; the consequent abolition of the domestic circle, and the perversion of all the social obligations which should exist between parent and child on the one hand, and between children themselves on the other.

The age at which a child became useful to its

parents, so long as the great mass of manufacturing was manual and confined to private dwelling-houses, was from fourteen to sixteen. At an earlier age it was useful in a minor degree as winder, &c.; but that was the period at which it became an auxiliary to the incomings of the family by working at the loom.

Before this it was a mere child, entirely dependent upon the exertions of its parents or older brothers and sisters for support. During this time it was taught, by daily experience, habits of subordination to its seniors. The period at which it ranked itself by the side of the efficient portions of the household, was a happy medium between too early an application and too late a procrastination of its physical energies; for the child was sufficiently matured, in its material organization, to bear without injury moderate and continued exertion; and no time had, as yet, been allowed for the acquirement of slothful habits: it came, too, at a time when the first impulses of puberty were beginning to stir new associations in his mind. These it checked by keeping him occupied, while he was removed from the influence of bad example, and laboured in an open workshop, free from the stimulus of warmth, and in the presence of his sisters, brothers, and parents,—the very best anodyne for allaying and keeping in due restraint his nascent passions,—whilst his moral and social instincts were under a process of incessant cultivation.

The same observations apply with still greater force to the females of the family. With them labour commenced at a somewhat earlier period, or they supplied the place of their mother in the household offices, leaving her at liberty to work for their sustenance, if such a course of proceeding was deemed necessary, or forced upon them by the pressure of circumstances. Whichever it was, she was kept from promiscuous intercourse with the other sex, at an age when it was to her of the utmost importance—her young sensibilities rendering her peculiarly liable to powerful and irresistible impressions.* It is true that the sports of her own sex were to some extent libidinous; but to these she was not admitted till a much later period; and these, though coarse and highly objectionable, rarely ended in mischief, being considered as a sort of prelude to marriage, universally existing amongst them—a custom most certainly "more honoured in the breach than in the observance."

Her occupations and feelings were therefore exclusively home-bred, and no idea existed that distinct or detached interests could intervene betwixt her parents and herself.

It is in these respects that the family of the factory labourers offers such strong contrasts and unhappy differences.

In the first place, there is no home labour. It

* Vide Introductory sketch—passim.

becomes therefore a mere shelter, in which their meals are hastily swallowed, and which offers them repose for the night. It has no endearing recollections which bind it on their memories—no hold upon their imaginations.

In the next place, the various members not only do not labour under their own roof, but they do not labour in common, neither in one mill; or if in one mill, so separated, that they have no opportunity of exchanging a single glance or a word throughout the long hours they are engaged there. Children are thus entirely removed from parental guardianship;—and not only so, but they are brought into immediate contact with parties, generally of their own age, equally removed with themselves from inspection, and equally unchecked by a consciousness that the eye of a brother, sister, or parent may be fixed upon them. They are placed, too, under the control of an overlooker, who from a sense of duty to his employer, if aggravated by no baser feeling, treats them frequently with harshness, often with brutal coarseness, making no allowance for childish simplicity, bashfulness, delicacy, or female failings; and this is most fatal to self-esteem,—for nothing so soon injures or destroys this, as unworthy treatment, suffered in themselves or witnessed in others, without the power of redress or even of appeal.

Again, they are subjected on all sides to the influence of vicious examples—in an heated atmosphere, and have no occupation, save watching the passage of a thread or the revolution of a spindle. The mind is but little engaged, there is no variety for it to feed upon,—it has none of the pure excitements which home affords,—it becomes crowded with images of the very opposite quality, and has its delicacy utterly and irretrievably ruined, and no opportunity is given for the growth of modesty on the one hand, or of the social obligations of brother, sister, or child on the other.

The next evil which removes factory labour another step still more widely apart from the condition of domestic manufacture is, that the wages of children have become, either by universal consent, or by the growth of disobedience, payable to the person earning them. This has led to another crying and grievous misfortune; namely, that each child ceases to view itself as a subordinate agent in the household; so far indeed loses the character and bearing of a child, that it pays over to its natural protector a stated sum for food and lodging; thus detaching itself from parental subjection and control. The members, therefore, of a spinner's or weaver's family become a body of distinct individuals, occupying occasionally, but by no means universally, the same home, each paying its quota to the joint expenses, and considering themselves as lodgers merely, and appropriating any surplus which may remain of their wages to their own private purposes, accountable to no one for the mode in which it happens to be used or wasted.

It is to be feared, that the mischiefs resulting from such an unnatural arrangement, must, in the first instance, be saddled upon the errors of parents—such a dereliction from filial duty being hardly likely to happen spontaneously on the side of the children ; and that a plan originally adopted in a few cases, by the family of idle and depraved parents,*—and many such are to be found, who would willingly batten upon the toil of their children—has become general, in consequence of the lowering in the reciprocal confidence and affection which ought to exist between parent and child.

In numerous examples then, at the present day, parents are thus become the keepers of lodging-houses for their offspring, between whom little intercourse beyond that relating to pecuniary profit and loss is carried on. In a vast number of others, children have been entirely driven away from their homes, either by unnatural treatment, or have voluntarily deserted them, and taken up their abode in other asylums, for the sake of saving a small sum in the amount of payment required for food and house-room.

This disruption of all the ties of home, is one of the most fatal consequences of the factory system.

* Too frequently the father, enjoying perfect health, and ample opportunities of employment, is supported in idleness on the earnings of his oppressed children.—Dr. Kay's pamphlet, p. 64.

The social relations which should distinguish the members of the same family, are destroyed. The domestic virtues—man's natural instincts, and the affections of the heart, are deadened and lost. Those feelings and actions which should be the charm of the fire-side—which should prepare young men and young women for fulfilling the duties of parents, are displaced by a selfishness utterly repugnant to all such sacred obligations. Tenderness of manner—solicitude during sickness—the foregoing of personal gratification for the sake of others--submission to home restraint—all these are lost, and their place occupied by individual independence—private avarice—the withholding assistance,* however slight, from those around them, who have a natural claim upon their generosity—calculations and arrangements, based solely upon pecuniary matters,—with a gradual extinction of those sympathies and feelings, which are alone fitted to afford happiness—a wearing away of the more delicate shades of character, which render home a world of pleasures, leaving nothing but attention to the simple wants of nature, in addition to the depraved appetites which are the result of other circumstances connected with their condition;—and in the end reducing them, as a mass, to a heartless assemblage of separate and

* "When age and decrepitude cripple the energies of the parents, their adult children abandon them to the scanty maintenance derived from parochial relief."--Dr. Kay's pamphlet, p. 64.

conflicting individuals, each striving for their "own hand," uninfluenced, unmodified by the more gentle, the more noble, and the more humanized cares, aspirations and feelings, which could alone render them estimable as fathers, mothers, brothers and sisters.

When it is borne in mind, at what an early period of life this separation of family takes place, its effects will be better and more correctly appreciated, and the permanence of the injurious impression produced by it, will be more clearly comprehended.

Factory labour, in many of its processes, requires little else but manual dexterity, and no physical strength; neither is there any thing for the mind to do in it; so that children, whose fingers are taught to move with great facility and rapidity, have all the requisites for it. Hence one reasonfor introducing mere infants into mills, though this is by no means the only one; and were the hours of labour sufficiently limited, and under proper regulation, when the present habits of their parents are considered, the evil—great in some respects as it is—would almost cease to be one. Children from nine to twelve years of age, are now become part of the staple hands, and are consequently subjected at this tender period to all the mischiefs incident to the condition of the older work-people.

It may be urged, that the mind of a child at this

age, cannot, from its very structure and previous impressions, be susceptible of the more vicious and immoral parts of the system; and that its previous education, which it is presumed must have been home, will, to some extent, guard it against evil communications.

It has been truly observed, and not the less beautifully than truly, that "heaven is around us in our infancy." This might have been extended, and said, that "heaven is around and within us in our infancy;" for the happiness of childhood springs full as much from an internal consciousness of delight, as from the novelty of its impressions from without. Its mind, providing the passions are properly guided, is indeed a swelling fountain of all that is beautiful—all that is amiable; —overflowing with joy and tenderness; and its young heart is a living laboratory of love, formed to be profusely scattered on all around it.

The very copiousness of its sensations, however, prevents stability in their direction, if not carefully tended—and if its heart and mind have capabilities for exhibiting and lavishing the treasures of their awakening energies, they are, from their very immaturity, more easily warped and misdirected The hacknied quotation, "just as the twig is bent," &c., is not the less true for being hacknied. Most unhappily, every thing which goes on before the eyes of the unfortunate factory children, is but too well calculated to nip in the bud—to wither in the spring time of its growth—the flower which was springing up within them, to adorn and beautify their future existence—and in its stead to bring forth an unsightly mass of ill-assorted and rugged excrescences, equally hateful to sight and injurious to the parent stem.

The independence assumed by older brothers and sisters, the total inattention to parental remonstrance or wishes, soon produce their influence upon a child—which is quite as ready to learn evil as good. Then, driven at the early age it is, into the mill, and at once placed amongst crowds of children similarly circumstanced with itself, the impressions made upon it at home soon become permanent. The subsequent possession of money, with the bickerings that arise therefrom, alienate any spark of affection which might still be lingering in its breast for its parents, and when a mere infant, it establishes itself either as an independent inmate of its paternal dwelling, or seeks out a lodging with other parties, as the case may happen to be.

The existence of a divided interest in a household, whether the division is between man and wife, or between parent and child, is alike fatal to its best interests. No home can ever be what it ought without proper government, or where all the inmates are on such terms of equality as to give to each an equal right to the direction of the whole—and even a household so constituted will not hang long together. In the homes of the manufacturing population, the divisions between parents and children, arising from the assumption of managing their own earnings, so generally acknowledged amongst them, deprives them of the most valuable portion of their influence.

Thus, whether at home or abroad, unfettered by wholesome restraints, the factory child grows up, acquiring vices of all shades, and utterly losing that which might render its condition one of respectability and comfort—the social and domestic virtues. Year after year rolls on, unfitting it more and more for the best purposes of life, and if it should become a parent, it transmits to its offspring the evils of a system of which it has been the victim.

The entire breaking up of households, which is an inevitable consequence of mill-labour, as it now exists, is one which may be regarded as the most powerfully demoralizing agent attendant upon it. This is, however, aided by many other causes, some of which have been already described, and others will be noticed in the course of the work.* The domestic affections, if they are to assume strength, must be steadily cultivated, and cultivated, too, in the only way of which they are capable. This is by parental kindness in the first place, which, by rendering home pleasant, and weaving its delightful associations in

* Vide the foregoing and subsequent chapters—passim.

the young imagination, forms one of the most sacred, most delightful, and most permanent feelings of the human heart. In the next place, the example of proper household subordination, for without this the first will be destroyed, or so weakened as to be inefficient and inoperative. Neither of these agents are brought to bear actively and properly upon the factory child. From its birth it sees nothing around it but dissension; its infant cries are hushed, not by maternal tenderness, but by doses of gin or opiates, or it is left to wail itself asleep from exhaustion. In thousands of cases it is abandoned throughout the day by its parents, both of whom are egaged in the mill, and left to the care of a stranger or a mere child*—badly used—badly-fed—its little heart hardened by harshness even in the cradle; then badly clothed—unattended during its growth by regular and systematic kindness—constantly hearing execrations, curses, blasphemy, and every thing coarse and obscene in expression—seeing on all sides strife, drunkenness, beastiality, and abominations, and finally sent shivering into the mill, to swell the hordes of children which have been similarly educated, and similarly abandoned to their own resources.

It may be said, that the agricultural labourer is subjected to a separation from his family, and that the members of his family are also, after a time, separated from home This is granted, and that, thus, *primâ facie*, he appears circumstanced in these respects like the factory labourer. Nothing can, however, be more dissimilar than the two cases, when looked at in their true bearings.

The agricultural labourer, it is true, pursues his occupation from home—but he pursues it in nine cases out of ten solitarily, or if he works in company, it is in small gangs; he works too in an atmosphere natural in its temperature, and favourable to bodily health, saying nothing of the moral influences of the sights and sounds which are his familiar companions; his labour is physically severe, and is just sufficient to require what intellectual capacity he generally possesses; his diet is plain and wholesome; he is freed from the example of many vices, by his situation, to which the factory labourer is exposed, and his habits and modes of life are simpler and purer. His family, separated from each other, and from home after a time, remain long enough under the paternal roof to have acquired some notion of domestic discipline, and that too under the best of all possible teachers—a mother, whose avocations are exclusively household. The labour of the sons, when old enough to pursue it, which is not till sixteen years of age, is that of the father, under similar circumstances. The daughters become household servants, either to persons of their own class, or, what is more general, in the houses of respectable families in the neighbourhood, or seek service in the surrounding towns and villages;* their family interest thus, of course, merging in that of their employers. In all these cases a strict watch is kept over their morals.† No point of similarity exists then between them, except in the single one of separation of families, and that too at a period and in a way to be as little injurious as possible to the moral character of the parties.

The agricultural labourer has other moral advantages over those possessed by the manufacturing one. He is frequently under the direct inspection of his employer, in the middle class of land proprietors, or respectable land-holders; and in the inferior order of both these, he is the personal assistant, and works in conjunction with it. In the highest order, he has the reflected benefit of hereditary rank and wealth, circumstances of more importance than the superficial observer is aware of, but which are rendered sufficiently apparent by examining into the condition of the cottagers and labourers upon those estates which

* Nine-tenths of domestic female servants, both in the metropolis and in all large towns, come from agricultural districts. So strong is the prejudice existing against town-bred servants, that many families absolutely refuse to take them under any circumstances.

† "It may be safely affirmed, that the virtue of female chastit does not exist amongst the lower orders of England, except to a certain extent among domestic female servants, who know that they hold their situations by that tenure, and are more prudent in consequence."—*Report of Poor Laws' Commission.*

are benefited by the residence of their proprietors.

These are a few of the moral advantages which he possesses still, to some extent, over the factory labourer. Of late years, indeed, the breaking-up of small farms—and other causes, have brought into operation upon him, the demoralizing agency of poverty and want of employ, and its influence has done much to deprive him of many of the benefits he once enjoyed.

In addition to the enumeration already given of the evils which result from the division of families, and the early age at which children are impressed into earning their own support, with the moral degradation which is their universal effect, another misfortune, of a very prominent character, attends upon the female division of the manufacturing population. This is, the entire want of instruction or example in learning the plainest elements of domestic economy; and this single circumstance goes far to explain many of the improvident habits which form a chief part of the curse upon their social condition. No earnings, however liberal, can compensate for this. It at once robs the home of the labouring man of every chance of being rightly or even decently conducted. If minute economy, which is the only true economy, is to be of service, it must be carefully taught, and with the best means of furnishing the supplies of a family, and making these supplies go to their utmost length. Of all these essentials to the head of a household, she is utterly ignorant, and her arrangements, if arrangements they can be called, where every thing is left to chance, are characterised by sluttish waste, negligence, carelessness as to the quality of food, and indifference as to the mode of cooking, and an absence of all that tidiness, cleanliness, and forethought which are requisite to a good housewife.

So complete is the separation of families, and so entirely are all their members absorbed by mill labour, that it very frequently happens that man and wife do not meet during the day at all. Working at different mills, perhaps at opposite sides of the town, their various meals are procured at some lodging-house in the immediate neighbourhood—thus adding another evil—another cause of the dissolution of the domestic links,—to the long list already brought under review.

CHAPTER IV.

SOCIAL SYSTEM—DOMESTIC HABITS, &c. &c.

Social Confederacy—Mode of life pursued by the Factory Labourer—Hour of commencing Work—Breakfast—Dinner—Tea, or Baggin—Their Constituents, &c.—Diet of the Manufacturing Population—Use of Tobacco—State of Markets—Saturday's Market—Description of the Morning and Evening—Effects of want of solidity in Diet—Dram-drinking—Increase of Gin vaults and Beer houses—Description of a Gin vault—Visit of the Male Labourer to the Gin vault—Visit of his Wife and Child—Girls and Young Women, their Visit to the Beer house—Conduct of Wives—Pawnbrokers' shops—The Use made of these by the Operative—Illicit Distilleries—Their Extent—Their localities described—Irish Population.

THERE is nothing which so truly marks the character of a community, in a moral point of view, as domestic manners; nothing which affords so correct and decided a criterion by which a judgment may be formed of its happiness and comfort. Politically speaking, the common people may be a dead letter, whilst their homes exhibit private independence and social enjoyment. Politically speaking, a people may possess many immunities—many rights—may even exercise a very marked control over the actions of their rulers, whilst their homes exhibit social disorganization and moral worthlessness.

The social confederacy of the present generation is full of anomalies. Possessing, as the great bulk of the population does, many advantages never known or dreamt of by their forefathers; education rapidly progressing; its wants liberally relieved; its sicknesses carefully tended; religion afforded it, nay, even brought to its doors, and applied to its senses; a practicability of earning something towards a livelihood; continual accessions of political privileges—it is nevertheless filled with immorality, irreligion, improvidence, political discontent, refusal to earn anything, ingratitude, ignorance, and vice, in every conceivable form in which it can develop itself.

Neither are these evils confined to one class of the labouring commmunity, proving very sufficiently that other causes must be at work beyond those dependent upon manufacture, on the one hand, and agriculture on the other. Neither is the excess of the existing demoralization less in the agricultural than in the commercial districts, though the one is a scattered population, and the other is gathered together in towns or crowded localities, circumstances in themselves unfavourable to health and morals; the former of which are consequently freed from many of those causes of declension which powerfully influence the latter.

The mode of life which the system of labour pursued in manufactories forces upon the operative, is one singularly unfavourable to domesticity. Rising at or before day-break, between four and five o'clock the year round, scarcely refreshed by his night's repose, he swallows a hasty meal, or hurries to the mill without taking any food whatever. At eight o'clock half an hour, and in some instances forty minutes, are allowed for breakfast. In many cases, the engine continues at work during mealtime, obliging the labourer to eat and still overlook his work. This, however, is not universal. This meal is brought to the mill, and generally consists of weak tea, of course nearly cold, with a little bread; in other instances, of milk-and-meal porridge. Tea, however, may be called the universal breakfast, flavoured of late years too often with gin or other stimulants. Where the hands live in immediate proximity to the mill, they visit home; but this rarely happens, as they are collected from all parts, some far, some near; but the majority too remote to leave the mill for that purpose. After this he is incessantly engaged—not a single minute of rest or relaxation being allowed him.

At twelve o'clock the engine stops, and an hour is given for dinner. The hands leave the mill, and seek their homes, where this meal is usually taken. It consists of potatoes boiled; very often eaten alone; sometimes with a little bacon, and sometimes with a portion of animal food. This latter is, however, only found at the tables of the more provident and reputable workmen. If as it often happens, the majority of the labourers reside at some distance, a great portion of the allotted time is necessarily taken up by the walk, or rather run, backwards and forwards. No time is allowed for the observances of ceremony. The meal has been imperfectly cooked, by some one left for that purpose, not unusually a mere child, or superannuated man or woman. The entire family surround the table, if they possess one, each striving which can most rapidly devour the miserable fare before them, which is sufficient, by its quantity, to satisfy the cravings of hunger, but possesses little nutritive quality. It is not half masticated; is hastily swallowed in crude morsels, and thrust into the stomach in a state unfavourable to the progress of those subsequent changes which it ought to undergo. As soon as this is effected, the family is again scattered. No rest has been taken; and even the exercise, such as it is, is useless, from its excess, and even harmful, being taken at a time when repose is necessary for the digestive operations.

Again they are closely immured from one o'clock till eight or nine, with the exception of twenty minutes, this being allowed for tea, or baggin-time, as it is called. This imperfect meal is almost universally taken in the mill: it consists of tea and wheaten bread, with very few exceptions. During the whole of this long period they are actively and unremittingly engaged in a crowded room and an elevated temperature, so that, when finally dismissed for the day, they are exhausted equally in body and mind.

It must be remembered, that father, mother, son, and daughter, are alike engaged; no one capable of working is spared to make home (to which, after a day of such toil and privation, they are hastening) comfortable and desirable. No clean and tidy wife appears to welcome her husband — no smiling and affectionate mother to receive her children—no home, cheerful and inviting, to make it regarded. On the contrary, all assemble there equally jaded; it is miserably furnished—dirty and squalid in its appearance. Another meal, sometimes of a better quality, is now taken, and they either seek that repose which is so much needed, or leave home in the pursuit of pleasure or amusements, which still farther tend to increase the evils under which they unavoidably labour.

The staple diet of the manufacturing population is potatoes and wheaten bread, washed down by tea or coffee.* Milk is but little used. Meal is consumed to some extent, either baked into cakes

* The increased consumption of tea and coffee, as compared to the increase in population, shows the change in the habits which has been going on so rapidly during the present century. The quantity of sugar consumed in 1814 was 1,997,000lbs.; in 1832, 3,655,000 lbs., an increase of 83 per cent.; increase in population, 24 per cent.; tea, in 1814, 19,224,000 lbs.; 1832, 31,548,000lbs, increase 65 per cent.; coffee, in 1814, 6,324,000lbs; 1832, 22,952,000lbs, increase 183 per cent., against an addition of population amounting to 24 per cent.

or boiled up with water, making a porridge at once nutritious, easy of digestion, and readily cooked. Animal food forms a very small part of their diet, and that which is eaten is often of an inferior quality. In the class of fine spinners and others, whose wages are very liberal, flesh meat is frequently added to their meals. Fish is bought to some extent, though by no means very largely; and even this not till it has undergone slight decomposition, having been first exposed in the markets, and, being unsaleable, is then hawked about the back streets and alleys, where it is disposed of for a mere trifle. Herrings are eaten not unusually; and though giving a relish to their otherways tasteless food, are not very well fitted for their use. The process of salting, which hardens the animal fibre, renders it difficult of digestion, dissolving slowly, and their stomachs do not possess the most active or energetic character. Eggs, too, form some portion of the operatives' diet. The staple, however, is tea and bread. Little trouble is required in preparing them for use; and this circumstance, joined to the want of proper domestic arrangements, favours their extensive use amongst a class so improvident and careless as the operative manufacturers.

Tobacco is very largely consumed by the male and female labourers indiscriminately; hundreds of men and women may be daily seen inhaling the fumes of this extraordinary plant, by means of short and blackened pipes.* Smoking, too, is an almost universal accompaniment to drinking—a pernicious habit, as will be shortly seen, prevailing to a frighłful extent in this portion of the population.

The difference exhibited both by the buyers and sellers of animal and vegetable food on the Saturday, which is the general provision market-day and the pay-day of the labourers, in the morning and evening, very strikingly illustrates the different grades into which the community of a manufacturing town or district is divided.

Speaking generally, the markets are well supplied, both as regards quality and quantity. Animal food, consisting of beef, mutton, veal, and pork, is plentiful, and of the best sorts—certainly not surpassed by any market in the kingdom. Vegetables are equally abundant and of an equally good quality. Lancashire has indeed long been famous for the excellence of its potatoes, a native rarely meeting with any fit to be compared with them for growth and flavour. Cheese, flour, butter, &c. &c., are also in like manner good and abundant—in short, there is nothing eatable but what may be found at a moderate price and in any quantity.

---

* The consumption of tobacco has increased from 1814 to 1832, from 15,000,000lbs. to 20,000,000 lbs, that is, about 31 per cent. against an increase in population of 24 per cent. It is much less used now than formerly by the upper and middle classes.

In the morning the markets are crowded with well-dressed respectable persons making their purchases for the ensuing week—order, civility, decency being preserved as far as these things can be on such occasions. The best animal food, whether it be flesh, fish, or fowl, is of course first carried away to the larders and cellars of the middle and upper classes, and the same with the vegetables. The day wears on, and about noon a change is observable in the appearance of the markets. The morning's trading had pretty well cleared them, a tolerably accurate balance being preserved between the supply of prime and first-rate articles, and the probable demand. At this time they are beginning to fill again—the butchers' stalls are replenished—the vegetables start into being—the sides of the streets in the principal market thoroughfares become lined with baskets or petty stalls, the property of sellers of every variety of minor article likely to tempt the cupidity or taste of vulgarity.

It might be supposed that a supply of similar quality to that of the morning was now to be found,—but it is widely different. Coarse—badly fed—too long kept, and not unfrequently diseased animal food, stands in the room of the excellent article of the morning. The vegetables have undergone an equal deterioration, consisting of the refuse of the morning's supply. These are retailed out, not by the respectable dealer or grower, but by a congregation of small buyers who have selected their stocks, not from their excellence, but from their cheapness;—and the same inferiority holds throughout. And what a scene is Saturday night's market—what a hubbub of discordant sounds—what jangling, swearing, drunkenness, noisy vociferation, confusion worse confounded, riot and debauchery. Thus passes Saturday till near midnight—a scene of turmoil, strife, and roguery.

If the perfection of social and domestic life consisted in limiting and stinting the supplies of man's natural wants, as to food—or if this perfection consisted less in the limiting these supplies, than in an indifference as to their quality,—a very great portion of the lowest classes in the great manufacturing towns are rapidly approaching, or rather have already nearly approached perfection. If true wisdom, as to eating, consists in simply satisfying the cravings of appetite, without reference to the nature or place of doing this—then do these people exhibit a high order of wisdom. There is, however, an intimate relation between moral and domestic virtues and modes of living. The Irish hand-loom weaver, who rarely tastes any food but potatoes, has reduced his scale of living to its utmost simplicity, and he holds the lowest rank in the very low classes even in these situations. Recklessness and improvidence may be ever detected by coarse, inferior, and badly cooked diet. Poverty, even in its very extremity, if still retaining any trace of self-respect, any tincture of a wish or hope for better things, will have its meals, even though they may be hardly worthy the name, with a regard to common decency and decorum. The savage who feeds promiscuously upon whatever comes before him, from his own species to ants and caterpillars, affords by this very omnivorousness, the most decisive proofs of his want of civilization, and domestic forethought and economy. So does the workman, in a different order of society, who consumes the refuse merely of those around him, without regard to its quality,—he proclaims, if possible, in still more decided language, the extent of his moral and social debasement.

The mere supply of the wants of nature, with respect to food, absorbs but a very small amount of the wages earned by the majority of the manufacturing labourers—confined as the supply is to the coarsest and most simple viands. Men may be found who have not yielded to the indifference and destructive habits of the mass, who are living in comfort and decency upon the average amount of wages earned by the whole population— a sufficient proof, if proof were wanting, that the mischief lies full as much with the labourers themselves, as in the system of labour;—bad as that is acknowledged to be.

The extinction of decent pride in their household establishments, which at present characterizes the mass of the manufacturing population, presents them in a very unfavourable point of view.

There are none of the minor comforts of existence—nothing but a hut of squalor and filth, alike repulsive to sight and smell and injurious to health—having few of the requisites of home, except as a place of mere shelter. It is stripped of every thing which might render it pleasant or delightful, and has in consequence no hold upon the affections. The labourer leaves it without regret—he anticipates no joy on his return—he finds there nothing but want—and all these are in a great measure the results, not of the pressure of actual want, but of his own bad and improvident habits.

The pure and quiet joys of home, may indeed never have been known either to himself or his family; but surely there is within the human heart—even within their hearts, hardened and debased as they are—some yearning after domestic bliss—some faint glimmer of a better and purer order of things—some longing to shake off a condition of life, that, like an incubus, presses upon and destroys all the best energies of their nature.

The operative having no home which can cheer the brief period allowed him from labour—destitute of moral principle, unguided and uninfluenced by good example—flies for relief to the gin-vault or the beer-house, dissipating in these haunts of crime and depravity, resources which, if properly applied, would furnish his house decently, supply his table with wholesome and nutritious food, and provide him with ample means to make him a respectable member of society.

The plainness and want of solidity and proper stimulus in the food of the labourer, is attended by some other evils, bearing strongly upon his domestic habits.

His labour is continued so uninterruptedly, that whether it is morning, or noon, or night, he leaves the mill or work-shop, and devours his watery meal with feelings of such mental depression and bodily exhaustion that he eagerly swallows a stimulus in the shape of spirits or beer, to supply by its temporary exciting influence the want of proper food on the one hand, and of due relaxation on the other.

This habit of dram-drinking, so fatal in its consequences, is of the most extensive prevalence. By satisfying the cravings for support, and by rousing into activity the mental faculties, the labourers, male as well as female, swallow the pernicious draught, and bless it as the boon which relieves them from their harassed sensations. They resemble in this respect the hypochondriac, who flies madly to a stimulant which in his better senses he deprecates and avoids as a curse; but there is this lamentable difference—that the labourers have no lucid intervals, no return of correct sensations. Day after day their toil is accumulated upon them. Deprived of the cheering influence of the face of Nature, robbed of the pure breath of heaven, cooped up in crowded buildings, with the Babel-like sounds of their companions, animate and inanimate, their overstrained minds and bodies know no return to healthy feelings, and they plunge deeper and deeper into the whirlpool, till they neither know their danger, nor, if they did, could they avoid or escape it, without a moral discipline, a physical regeneration, which at present appear, if not utterly hopeless, at least very remote.

The increase which has of late years taken place in the number of gin-vaults, and the more than equal increase in the number of low beer-houses since the passing of the bill termed the Beer Bill,—though its more correct designation would have been a "bill for the demoralization of the working classes"—is sufficiently indicative of the prevalence of dram-drinking and tavern-haunting.*

In Manchester alone there are very near if not quite one thousand inns, beer-houses, and gin-vaults. Of these more than nine-tenths are kept open exclusively for the supply of the labouring population,† placed in situations calculated for

---

* Mr. Braidley, the respectable and intelligent boroughreeve of Manchester during 1832-3, observed the number of persons entering a gin shop, in five minutes, during eight successive Saturday nights, and at different periods, from seven o'clock till ten. The average result was 112 men and 163 women, or 275 in forty minutes, which is equal to 412 per hour.—Dr. Kay's pamphlet, p. 58.

† Manchester is divided into districts for municipal convenience. Of these Nos. 1, 2, 3, 4, belong exclusively to the labouring population, including with these, Nos. 5 and 6, there are, in these localities, 270 taverns, 216 gin shops, and 188 beer houses—total, 674, which minister almost entirely to the wants of the poor.—Dr. Kay's pamphlet, p. 58.

their convenience, decked out with every thing that can allure them, crowded into back streets and alleys, or flaunting with the most gaudy and expensive decorations in the great working thoroughfares. They are open at the earliest hour, when the shivering artizan is proceeding to his work, holding out to him a temptation utterly irresistible—and remain open during a considerable portion of the night ministering their poisons to thousands of debilitated creatures, who flock to them, in place of seeking excitement and pleasurable stimulus in fire-side comforts and enjoyments.

Nor is it the adult male labourer who alone visits these receptacles for every thing that is wicked and degraded. Alas! no. The mother with her wailing child, the girl in company with her sweetheart, the mother in company with her daughter, the father with his son, the grey-haired grandsire with his half-clad grand-child, all come here—herding promiscuously with prostitutes, pickpockets, the very scum and refuse of society—all jumbled together in an heterogeneous mass of evil, to the ruin of every thing chaste and delicate in woman, and the utter annihilation of all honourable or honest feeling in man.

Thus mingling in wild carouse, crimes of all shades are perpetrated. Blasphemy, fornication, adultery, incest, child-murder, form the black dark back ground; while drunkenness, thieving, and obscenity stand out boldly in front. The very sources of every noble principle of the human heart are depraved, and converted into pollution. The mind loses its healthy tone, and remains dormant except when under the influence of these abominable stimulants—whilst the body becomes emaciated, shrunk, dwindled to a mere anatomy, and sinks into premature decrepitude, wearing out its miserable remnant of existence in the workhouse—or becomes the victim to some acute disease and dying in the crowded wards of a hospital—falls into the grave unheeded, uncared for even by those who have derived their very existence from the blasted trunk, now withering before them.

It is a strange sight to watch one of these dens of wickedness throughout an evening: it is a strange, a melancholy, yet, to the meditative man, an interesting sight. There approaches a half-clad man, covered with cardings, shivering even beneath the summer breeze which is singing around him. He comes with faltering step, downcast eye, and air of general exhaustion and dejection. He reaches his accustomed gin-vault, disappears for half an hour or less,—and now comes forth a new creature: were it not for his filthy dress, he would hardly be recognized—for his step is elastic, his eye is brilliant and open, his air animated and joyous. He inhales the breeze as a refreshing draught, and he deems himself happy. His enjoyment is, however, short-lived, and purchased at an immense sacrifice, for the

> "—Price is death!
> It is a costly feast."

Now comes a woman, perhaps his wife, bearing a sickly and cadaverous-looking infant, wailing and moaning as if in pain or wanting nutriment. She is indeed offering it the breast, but it is flaccid and cold as marble. She has no endearments for her child, it is held as a burden—passively and carelessly. She is thin, pale, and badly dressed—is without bonnet, and her cap is soiled and ragged; her bosom is exposed, her gown is filthy, her shoes only half on her feet, and her whole aspect forlorn and forbidding. She too disappears for a time within the gin-shop, remains longer than her husband, but returns equally changed. The child is now crowing in her arms, clapping its tiny hands, and is filled with infantine mirth,—whilst its mother views it with fondness, joins in its vociferations, tosses it in her arms and kisses it like a mother. She passes on cheerily, her whole gait is altered, her cheeks are flushed, and she thinks herself happy—for her maternal feelings are aroused, and her inebriated child seems to her own disordered senses the very paragon of beauty and delight.

The pair have now reached home—night is far advanced, and the fumes of their intoxications are worn off or become converted into sullenness. The child is in a stupor, and the husband and wife meet without a single kindly greeting. There is no food, no fire: bickerings arise, mutual recrimination, blows, curses—till both at last sink into the stupified sleep of drunkenness, worn out by toil, excessive stimulus, and evil passions—leaving the child lying on a ricketty chair, from which it must inevitably fall should it awake.

Here come several girls and young women, tolerably dressed; some with harsh, husky voices, shewing the premature development of puberty, others full grown and perfectly-formed women. All, save one, have the same pallid hue of countenance, the same coarseness of expression, the same contour of figure—but all seem equally toil-worn and exhausted. One amongst them is, however, beautiful, and beautiful as an innocent girl alone can be—the very purity of her heart and her soul gleaming in her face. Her figure is plump and round, and her cheeks, though somewhat pale, are yet firm in their outline. It is evident that she is scarcely at home in the presence of her companions, nor one of them in feeling, though it would seem that she is condemned to the same labour. Yes! it is so. She is not many weeks returned from a distant town, in which she had been apprenticed to a respectable trade. Adverse circumstances have, however, driven her home, and she has no resource but to become a weaver, and this she has been for upwards of a week. She hesitates to enter the beer shop—she withdraws timidly, but at length is lost within its door, amidst the laughter and jeers of her companions. They remain long; and now approach a number of young men with soiled dress, open necks, and of obscene speech. They, too, enter the beer house. Laughter long and loud resounds from it; time wears on, but the drunken revel continues unabated—now shewing itself by bursts of obstreperous merriment—now by vollies of imprecations—now by the rude dance—and now by the ribald song. But where is that delicate and beautiful girl? Can she be one sharing such scenes? Can she, whose eyes and ears evidently revolted from the bold gestures and speeches of her companions, be remaining to share such coarse orgies? Eleven o'clock, and the party re-appear. Cursing, swearing, hiccuping, indecent displays, mark their exit; and *there is* the fair girl, whose "unsmirched brow" so lately gave token of her purity. But now she is metamorphosed into a bacchanal, with distended and glowing cheeks, staggering step, disordered apparel—lost, utterly lost, to herself; and when the morning bell rings her to her appointed labour, she will be one of the herd, and will speedily lose all trace of her purity and feminine beauty.

If there is one period of the day when these displays excite more unmingled disgust than another, it is during the hours of labour, when those wives and mothers of the absent artizans, who are either unable to work, or cannot procure it, are left at home. No domestic cares occupy them, except in a few rare cases. Their hearths are unswept, their persons and houses uncleaned, their rooms untidied. On the contrary, they are seen in groups of two or three, lounging about the gin or beer shops, wasting any trifle of money they may have reserved, or procuring a supply by pawning some portion of their wretched ap- apparel, or equally wretched furniture.

Pawnbrokers' shops have an affinity in their demoralizing agency to the gin and beer houses, and are almost as numerous, occupying the same localities, and giving unhappy facilities to the poor man to protract his Saturday night's, or Sunday's debauch throughout the early part of the week. Article after article is pledged for small sums, to be redeemed the next pay-day, and homesteads which, on the Saturday mornings, are destitute of every domestic utensil, are found, in the evening, possessing some of these. They have been fetched from pledge, a loss has been sustained upon them, and in ninety-nine cases out of one hundred they are again missing before the middle of the ensuing week. The same with regard to dress—coat or trowsers, cloak, dress, or under dress, shoes, stockings, &c. &c., all have a constant round. Sacrifice after sacrifice takes place, till at length the articles are either left unredeemed, or are become so far worn and tattered as to be no longer a valid security to the pawn-broker, even for the very small sum he advances upon them, be it twopence, fourpence, sixpence, eightpence, or one shilling.

This utter disregard for personal comfort and household decency, strongly demonstrates the miserably low ebb to which the domestic relations are reduced, and proclaims, in a voice which it is impossible to misunderstand, how deplorable, how pitiable is the condition, which thus deprives itself, like the untutored and uncultivated savage, of every thing humanising, for the sake of a momentary excitement;—which leaves matters in an infinitely worse state than before, and precipitates the unhappy being who thus blindly indulges a depraved taste, only the more rapidly upon his inevitable ruin.* Truly did Mr. Hunter remark, that of all the products of civilization, the North American savage was anxious alone for brandy

* Neither has this passion lessened since the day of the acute Hunter. At the treaty of Chicago, in 1821, the commissioners ordered that no spirits should be issued to the Indians. A deputation of the chiefs was sent to remonstrate against this precautionary measure; and at its head was Topnibe, chief of the Potawatomic tribe, a man upwards of eighty years of age. Every argument was used to convince them that the measure was indispensable: that they were exposed to daily murders, and that while in a state of intoxication, they were unable to attend to the business for which they were convened. All this was useless, and discussion only terminated by the peremptory refusal of the commissioners to accede to their request. "Father," said the hoary-headed chief, when he was urged to remain sober, and make a good bargain for his people, "Father, we care not for money, nor the lands, nor the goods.—We want the whiskey. Give us the whiskey."

and gunpowder; and with equal truth it may be said, that the nearer a people approach to the destitution of savage life, the more eagerly do they seek to participate the enjoyments of intoxication. Distilled spirits afford a cheap and effectual oblivion of cares and wants, and however much the indulgence may increase the sum of human sufferings, it is not surprising that they who have no other pleasure within their reach, should madly snatch at this solitary comfort.

Independently of these open and recognised means, which are within the the reach of the labourer and his family, there are others of most extensive operation for demoralizing him, through the agency of drunkenness and its attendant vices. The lowest class of the population, in the manufacturing towns, is, to a very considerable extent, made up of Irish, who inhabit the cellars and those portions of the towns which, to the casual observer, would appear totally uninhabitable. In these places, however, is congregated a very numerous body of people, who have introduced with them the same spirit of recklessness and improvidence, with the same systematic evasion of or violent resistance to law and order, which have so long disgraced their native country, and gone far towards ruining its resources. Here, and by these individuals, illicit distillation is carried on to a great extent. It has been calculated, upon not very perfect data, that there are not less than 100 stills in constant operation in Manchester alone,* producing genuine potheen of the highest strength. From many seizures which have been made, it appears that the average size of the tubs may be about thirty gallons, and if each distillery should produce no greater quantity than this, the whole amount would be very considerable; and this is drunk, be it remembered, almost exclusively by the lowest classes. Reckoning thirty gallons as the weekly produce, it would give annually 156,000 gallons, which pays no duty, and the manufacture of which is carried on under circumstances in the highest degree unfavourable to the welfare of a peaceable population. This estimate is, however, most probably underrated—the difficulty of detection is extreme—occupying, as the distilleries do, sites apparently the most improbable, and unfitted for the purpose. These generally are dark cellars, having no outlet except a trap-door, opening into some obscure court, half filled with filth, or excavations dug in the sand-stone rock, beneath tenements occupied by persons either knowing nothing of their subterranean neighbours, or being in league with them. The requisite apparatus is so simple and so little expensive, that

* During the year 1832, thirty persons were committed to the New Bailey, Manchester, for having either been detected in distilling or hawking illicit spirits. Nearly the whole of these were Irish—one half had been known to have carried on the trade before, and one-third had been previously convicted and suffered imprisonment.

detection is no hinderance, and a seizure valueless. The fine inflicted, £30, is of no moment to the successful speculator; and if it should so happen that he is unable to pay it, the short imprisonment which is the alternative, only hardens him in crime, and turns him out upon society a more determined and experienced scoundrel.

The introduction of a low Irish population into Manchester and the surrounding manufacturing towns and districts, has unquestionably aided very materially the destruction of domestic virtues and orderly habits in the operatives. The disregard to home comforts, which renders the Irish cabin a blot upon the history of its country, is exhibited still more strikingly when seen in the midst of a large town; and it has, unfortunately, found ready imitators in a class of the community, prepared for its adoption, in some degree, by poverty, ignorance, want of morality, and a growing indisposition for home, generated by a system of labour which, by separating families, and by exhausting their physical energies by incessant application, rendered them ready to fly, for temporary relief, to the gin-shop, the beer-house or the whiskey dealer.

CHAPTER V.

SOCIAL CONDITION—HABITATIONS—DOMESTIC HABITS, &c. &c.

Progress of Civilization—The Manufacturing Population—its Degradation—Homes—Account of Houses built for the Poor—Occupation by several Families of one House—Effects of this upon Decency and Morals—Language, its Obscenity—Cellar residences—Their numbers—Their Occupiers—Irish—Lodging houses—Wages devoted to Household Purposes—Dress—Maternal Affection—Love of Infancy in Woman—Destruction of maternal love by Factory Labour—Husband and Wife—Married love—Brother and Sister—Parent and Child—A Household described.

If domestic manners and modes of living are means by which a judgment may be formed of the condition of a community, its habitations, furniture, dress, &c. &c., are subsidiary aids of no mean value. It is true they form but one part of a system of social organization; but as developing traits of character, and shewing where evils are the greatest and the most conspicuous, important hints may be derived from them for amelioration, which amelioration, whether it is to result from legislative enactments, or from the efforts of private philanthropy, must be preceded by a clear understanding of the mischiefs they would amend, or their endeavours will do no good, but may, from want of proper direction, do infinite harm.

The inquirer into the progress of the civilization of man, has long ago learnt the following series of facts. He finds that as man removes from utter barbarism—from a state elevated but one remove above the brute creation—he raises himself a habitation, more or less comfortable, as a shelter against the vicissitudes of the seasons, and as a place of refuge; that in his primeval condition he contented himself with the shelter of a tree, and the protection of the natural caves or strong-holds around him; that, at this period, he lives upon aliment chiefly of a vegetable nature, or picks up a scanty and precarious subsistence by fishing, or devouring the minuter forms of animal life; that in his first advance in social improvement, he erects himself a rude hut of wood, or sods, or stone, as his particular locality may point out to him, badly built, badly covered-in, admitting little light, but freely open to the winds of heaven; that he now becomes a hunter, and tills imperfectly a small patch of ground, for the production of those esculent roots or seeds which experience has taught him are fit for the support of life; that in his next advance he improves his hut, forms a communion with his fellows under an acknowledged superior, joins his labours, whether of the chace or of cultivation, with those of others—shews some tokens of religion, however barbarous and superstitious—selects a woman as a companion, and lays the foundation of the relations of husband and wife; that in his continued advance, his cottage becomes to him more than a place of mere shelter; that its walls are now covered with the products of war or the chace; that it has other inmates—a family of children; that his moral attributes slowly and imperfectly develop themselves; that his labour or sport has for its aim the maintenance of his family; or that his wife, not yet freed from savage thraldom, is the principal agent in the production of food; that as he still progresses, his home is better built, assumes a different aspect; its interior having many simple decorations, and is neat and clean, whilst its outside bears marks of attention being paid to effect; that he now surrounds it with a patch of ground, over which he claims a right of exclusive possession; that his wife now becomes to him more than a creature retained solely for the gratification of his appetites, and that his children are looked upon as beings in whose welfare he is deeply and sensitively interested; that he submits to codes of laws, or municipal regulations, which, although they may interfere with his individual liberty or particular rights of property, are yet obviously beneficial to the interests of the community of which he forms a member; that he is now stationary, has lost his predatory habits, and has assumed his rank as a social and moral being; that in his further advances he still improves his habitation, builds his house in a more durable manner, and with better materials, divides it into distinct compartments, and separates the sexes; that his wife is no longer an instrument of labour, but depends upon him for support; that promiscuous intercourse between the sexes is condemned and prohibited as injurious to the marriage contract; and that thus, step after step, he goes on to the maximum of civilization and excellence of social confederation, exhibiting, in habitation, dress, and manners, a congruity, an homogeneousness of improvement, shewing how intimately all these separate conditions are essential to the perfection of the whole system.

If the progress of civilization is thus clearly marked by these various gradations from the simple animal existence of man in his primeval state, his lapse may be truly said to be indicated by data of a similar character. Taking the extent of refinement and the perfection of social communion as they are displayed by the middle class of society—neither placing it too high nor too low—the degree to which the lowest classes in the manufacturing towns and districts have retrograded, or remained behind in the march of improvement, is very apparent. Judging them by the same rules which have been applied to mark the advancement of man from a savage state they have made but few steps forward; and though their primitive nature is disguised and modified by the force of external circumstances, they differ but little in inherent qualities from the uncultivated child of nature, and shew their distinction rather in the mode than the reality of their barbarous and debased condition.

When it is borne in mind that the class which is so little elevated in its social instincts and domestic habits and intelligence, lives in the nineteenth century, in a country which has been long freed from hostile aggression, in the midst of a nation pre-eminent for its cultivation of the arts and sciences—celebrated for its benevolence, and its unceasing efforts to extend the blessings of religion and moral instruction over the whole habitable globe—famed for the general extent of its education—its enjoyments of political rights—its charitable institutions—the numbers of its clergy—the wealth and splendour of its church—its degraded condition becomes the more remarkable. Did this unhappy depression in the social rank include the population only of some secluded and out of the way nook of the empire; did it only embrace a few scattered tribes, remote from the centre of its power; did it only extend over a limited number of individuals which, compared to the great mass, were but as a mote in the sun-beam, it might excite but little surprise. But, no—it is just the reverse. It includes several millions in a small population—placed too in the very heart of the nation—important and indispensable agents in upholding its stability—nay, the very cornerstone of a great portion of its pre-eminence—

having the full benefit of all those institutions so lauded and pointed out as land-marks, shewing how far it has advanced in the march of civilization.

The houses of great numbers of the labouring community in the manufacturing districts present many of the traces of savage life. Filthy, unfurnished, deprived of all the accessories to decency or comfort, they are indeed but too truly an index of the vicious and depraved lives of their inmates. What little furniture is found in them is of the rudest and most common sort, and very often in fragments—one or two rush-bottomed chairs, a deal table, a few stools, broken earthenware, such as dishes, tea-cups, &c. &c., one or more tin kettles and cans, a few knives and forks, a piece of broken iron, serving as a poker, no fender, a bedstead or not, as the case may happen to be, blankets and sheets in the strict meaning of the words unknown—their place often being made up of sacking, a heap of flocks, or a bundle of straw, supplying the want of a proper bedstead and feather bed, and all these cooped in a single room, which serves as a place for domestic and household occupations.

In those divisions of the manufacturing towns occupied by the lower classes of inhabitants, whether engaged in mill-labour alone, or in mill-labour conjointly with hand-loom weaving, the houses are of the most flimsy and imperfect structure. Tenanted by the week by an improvident and changeable set of beings, the owners seldom lay out any money upon them, and seem indeed only anxious that they should be tenantable at all, long enough to reimburse them for the first outlay. Hence, in a very few years they become ruinous to a degree. One of the circumstances in which they are especially defective,* is that

* The following table, arranged by the classification committee of the Special Board of Health appointed during the late irruption of cholera into Manchester, affords the most decisive evidence upon this point:—

District.	Number of houses inspected	Houses requiring white-washing.	Houses out of repair.	Houses wanting proper soughing.	Houses damp.	Houses ill ventilated.	Houses wanting privies.
1	850	399	128	112	177	70	326
2	2489	898	282	145	497	109	755
3	213	145	104	41	61	52	96
4	650	279	106	105	134	69	250
5	413	176	82	70	101	11	66
6	12	3	5	5			5
7	343	76	59	57	86	21	79
8	132	35	30	39	48	22	20
9	128	34	32	24	39	19	25
10	370	195	53	123	54	2	232
11							
12	113	33	23	27	24	16	52
13	757	218	44	103	146	54	177
14	481	74	13	83	68	7	133
Total. .	6,951	2,565	960	939	1435	452	2,221

"These numerical results fail to exhibit a perfect picture of the ills which are suffered by the poor. The replies to the questions contained in the inspector's table, refer only to cases of the most positive kind, and the numerical results would therefore have been exceedingly increased, had they embraced those in which the evils existed in an inferior degree. Some idea of the want of cleanliness prevalent in their habitations, may be obtained from the report in the number of houses requiring whitewashing; but this column fails to indicate their gross neglect of order and absolute filth. Much less can we obtain satisfactory statistical results concerning the want of furniture, especially of bedding, and of food, clothing, and fuel.—Dr Kay's pamphlet, p. 31-2.

of drainage and water-closets. Whole ranges of these houses are either totally undrained, or only very partially soughed. The whole of the washings and filth from these consequently are thrown into the front or back street, which being often unpaved and cut up into deep ruts, allows them to collect into stinking and stagnant pools, while fifty, or more even than that number, having only a single convenience common to them all, it is in a very short time completely choked up with excrementitious matter. No alternative is left to the inhabitants but adding this to the already defiled street, and thus leading to a violation of all those decencies which shed a protection over family morals.*

* The subjoined table, the result of inquiries, made by the Special Board of Health in 1832, shows the state of the streets in Manchester, and shows how intimately localities and characters are connected:—

Districts.	Streets inspected.	Streets unpaved.	Streets in part paved.	Streets ill ventilated.	Streets containing heaps of refuse, stagnant pools, ordure, &c.
1	114	63	13	7	64
2	180	93	7	23	92
3	49	2	2	12	28
4	66	37	10	12	52
5	30	2	5	5	12
6	2	1	0	1	2
7	53	13	5	12	17
8	16	2	1	2	7
9	48	0	0	9	20
10	29	19	0	10	23
11	0	0	0	0	0
12	12	0	1	1	4
13	55	3	9	10	23
14	33	13	0	8	8
Total.	687	248	53	112	352

It very frequently happens that one tenement is held by several families, one room, or at most two, being generally looked upon as affording sufficient convenience for all the household purposes of four or five individuals. The demoralizing effects of this utter absence of social and domestic privacy must be seen before they can be thoroughly understoood, or their extent appreciated. By laying bare all the wants and actions of the sexes, it strips them of outward regard for decency—modesty is annihilated—the father and the mother, the brother and the sister, the male and female lodger, do not scruple to commit acts in the pre-

---

"An accurate inspection of this table, will render the extent of the evil affecting the poor more apparent. Those districts which are almost exclusively inhabited by the labouring population, are, Nos. 1, 2, 3, 4, and 10; Nos. 13, 14, and 7, also contain, besides the dwellings of the operatives, those of shopkeepers and tradesmen, and are traversed by many of the principal thoroughfares. No. 11, was not inspected; and Nos. 5, 6, 8, and 9, are the central districts containing the chief streets, the most respectable shops, the dwellings of the more wealthy inhabitants, and the warehouses of the merchants and manufacturers. Substracting, therefore, from the various totals, those items in the reports which concern these divisions only, we discover in those districts which contain a large portion of poor, *viz.* 1, 2, 3, 4, 7, 10, 13,14, that among 579 streets inspected, 243 were unpaved, 46 partly paved, 93 ill-ventilated, and 307 contained heaps of refuse, deep ruts, stagnant pools, ordure, &c.; and in the districts almost exclusively inhabited by the poor, *viz.* 1, 2, 3, 4, and 10, out of 438 streets inspected, 214 were unpaved, 32 partly paved, 63 ill-ventilated, and 259 contained heaps of refuse, ruts, stagnant pools, ordure, &c.—Dr. Kay's pamphlet, p. 30.

sence of each other, which even the savage hides from the eyes of his fellows.

The brutalizing agency of this mode of life is very strongly displayed in the language employed by the manufacturing population, young and old alike. Coarse and obscene expressions are their household words; indecent allusions are heard proceeding from the lips of brother to sister, and from sister to brother. The infant lisps words which, by common consent, are banished general society. Epithets are bandied from mother to child, and from child to mother, and between child and child, containing the grossest terms of indecency. Husband and wife address each other in a form of speech which would be disgraceful to a brothel—and these things may be imputed in a very considerable degree to the promiscuous way in which families herd together; a way that prevents all privacy, and which, by bringing into open day things which delicacy commands should be shrouded from observation, destroys all notions of sexual decency and domestic chastity.*

---

* "In addition to overt acts of vice, there is a coarseness and grossness of feeling, and an habitual indecency of conversation, which we would fain hope and believe are not the prevailing characteristics of our country. The effect of this upon the minds of the young will be readily conceived; and is it likely that any instruction or education, or Sunday schools, or sermons, can counteract the baneful influence, the insinuating virus, the putrefaction, the contagion of this moral depravity which reigns around them.

> Nil dictu visuque fædum hoc lumina tangat
> Intra quæ puer est.—JUVENAL."

—*Enquiry*, &c. p. 25.

It may be questioned, whether in any other quarter of the world, or in any other condition of society, such an absence of the observances of modesty and personal cleanliness can be found. Nothing can be more brutalizing—nothing can render individuals more debased in their feelings and habits—nothing can tend more powerfully to produce that personal coarseness of habits, and filthy indelicacy, which are so disgusting and repulsive in woman.

Many of these ranges of houses are built back to back, fronting one way into a narrow court, across which the inmates of the opposite houses may shake hands without stepping out of their own doors; and the other way, into a back street, unpaved and unsewered. Most of these houses have cellars beneath them, occupied—if it is possible to find a lower class—by a still lower class than those living above them. From some recent inquiries on the subject, it would appear, that upwards of 20,000 individuals live in cellars in Manchester alone. These are generally Irish families—hand-loom weavers, bricklayers' labourers, &c. &c., whose children are beggars or match-sellers in conjunction with their mothers. The crowds of beings that emerge from these dwellings every morning, are truly astonishing, and present very little variety as to respectability of appearance: all are ragged, all are filthy, all are squalid. They separate to pursue their various callings, either shutting up their dens till night, or leaving a child as sole occupant. A great portion of these wander about the town and its suburbs, begging or stealing, as the case may be; others hawk little matters, such as pins, matches, oranges, &c., bringing back with them any fragment of meat or bread they have been able to procure. These cells are the very picture of loathsomeness—placed upon the soil, though partly flagged, without drains, subjected to being occasionally overflowed, seldom cleaned—every return of their inmates bringing with it a farther accession of filth—they speedily become disgusting receptacles of every species of vermin which can infest the human body.

The domestic habits of these improvident creatures are vile in the extreme—carrying their want of household decency, if possible, one step further than those which have just been described. The Irish cottier has brought with him his disgusting domestic companion the pig; for whenever he can scrape together a sufficient sum for the purchase of one of these animals, it becomes an inmate of his cellar.*

* "In all respects the habitations of the Irish are most destitute. A whole family is often accommodated on a single bed, and sometimes a heap of filthy straw, and a covering of old sacking hides them in one undistinguished heap, debased alike by penury, want of economy, and dissolute habits. Frequently the inspectors found two more families crowded into one small house, containing only two apartments, one in which they slept, and another in which they ate; and often more than one family lived in a damp cellar, containing only one room, in

It is here too that he displays his recklessness in another of his characteristic propensities—whiskey-drinking, an opportunity for the indulgence of which is furnished by the illicit distillers in his vicinity for a mere trifle. The disgraceful riots which are calling perpetually for the interference of the police, are mainly attributable to this cause, and a return from the lock-ups would abundantly show how terrible are the outrages inflicted upon each other during these drunken brawls. Often, indeed, the whole population of court, street, or entire district, forms a faction, in opposition to that of some other in the neighbourhood; and the cries of "O'Flanagan" and "M'Carthy," are as rife as in the heart of Connaught. When their passions are roused by intoxication, most severe and often bloody conflicts ensue between them, to the disturbance and degradation of the more peaceable inhabitants. Thus it appears that the inferior order of Irishmen have brought with them all their vices into the manufacturing districts, and aid powerfully by their example—independently of lowering the value of the labour of the English operative—the demoralization which marks his general character.

Another fertile source of the licentiousness in

---

whose pestilential atmosphere from twelve to sixteen persons were crowded. To these fertile sources of disease were sometimes added the keeping of pigs and other animals in the house, with other nuisances of the most revolting character."—Dr. Kay's pamphlet, p. 32.

domestic manners, exists in the number of lodging-houses, which are very abundant in all the manufacturing districts. In towns they are thickly scattered through those divisions occupied by the poor. By a survey made in Manchester in 1832, there were found very near 300 of these houses. When it is remembered that these are but the temporary asylums of want and depravity, their number, great as it is, affords no criterion for ascertaining how many persons become their inmates during the year. In another point of view, they are extremely injurious: the breaking up of families, the consequence of mill labour, drives many of those who should be sheltered under a very different roof, to take up their abode in these haunts of crime; where, if not already debased by other causes, they are speedily reduced to the very lowest ebb of moral depravity. Their influence in this respect, resembles very closely that brought about by a residence in prison; for the parties which are the habitual occupants of the one, are in their turn found living in the other.

The extraordinary sights presented by these lodging-houses during the night, are deplorable in the extreme, and must fill the heart of any man, open to the feelings of humanity, with pain and unutterable disgust. Five, six, seven beds—according to the capacity of the rooms—are arranged on the floor—there being in the generality of cases, no bedsteads, or any substitutes for them; these are covered with clothing of the most scanty and filthy description. They are occupied indiscriminately by persons of both sexes, strangers perhaps to each other, except a few of the regular occupants. Young men and young women; men, wives, and their children—all lying in a noisome atmosphere, swarming with vermin, and often intoxicated. But a veil must be drawn over the atrocities which are committed: suffice it to say, that villany, debauchery and licentiousness, are here portrayed in their darkest character. They serve as so many foci for crime—so many hot-beds for bringing into existence vices which might have lain dormant, if not roused into vitality by their unnatural stimulus.

The small sum devoted to household purposes by the operative, may be determined with some accuracy, and it affords considerable information as to his habits. A family consisting of five persons may serve as an example, that being about the average number. This family will pay for rent, which includes taxes, 3*s.* a week for a cottage containing two rooms; and the different items of their expenditure will be somewhere as follows: tea, quarter of a pound, 1*s.* 2*d.*; coffee, half a pound, 10*d.*; bread, 3*s.*; coals and candles, 1*s.*; animal food, 2*s.*; butter, soap, salt and cheese, 2*s.* 6*d.*; potatoes, 1*s.*; sugar, 1*s.* 6*d.*; milk, 3*d.*;—total, 16*s.* 3*d.* The allowance of the above articles is liberal, and certainly more than what the generality of families use. Against this must be placed the amount of their earnings, which when taken at the average rate of wages paid to manufacturing labourers, will be 10*s*. 6*d*. per head—total of wages, 2*l*. 12*s*. 6*d*.; thus leaving a surplus of 1*l*. 16*s*. 3*d*. for dress and other purposes.

Though the general appearance of the operative is squalid, and the majority of the middle aged badly dressed, the young women and girls expend considerable sums upon their persons. They exhibit in the style of dress a very striking contrast with the inhabitants of the rural districts, a distinction always observable between town and country girls. A taste for shewy clothing has been no doubt given by the extreme cheapness of printed calicoes and muslins, and indeed by the lowness of price in all manufactured articles of middling quality; and the factory girl, with her pale face and languid expression, offers, when decked out in her Sunday and holiday apparel, a strange anomaly with the dirty unfurnished, and miserable home from which she issues.

The females employed in silk mills go a step still further as to dress than the cotton spinner or weaver. Many of these are really well and handsomely clad, and present an appearance of outward respectability, equal if not superior to that of hundreds in the middle walks of life.

A laxity in morals is, however, by no means incompatible with personal tidiness and neatness. This fact is sufficiently proved by the conduct and gestures of many of these girls. A disposition for display and a looseness of deportment may be too generally remarked in these victims of a vicious and hateful system, which has the power to convert all the tenderness of a woman's heart and the purity of her feelings into vanity and lasciviousness.

Maternal affection is one of those beautiful and beneficent instincts which so strongly mark the goodness and surpassing wisdom of the great Author of nature. It is alike influential upon the unwieldy whale, and the most minute mammiferous animal—upon the fierce and cruel eagle, and the painted humming-bird, living a life of sweets in the odoriferous regions of which it is a denizen. In all ages, in all countries, in all stages of civilization—in war, in pestilence, shipwreck, or famine—whether roaming through the jungle or over the prairie—whether traversing the expanse of the continent or dwelling in the far off and isolated island—woman has ever been found with the hallowed character of a mother, and exhibiting, for the sake and love of her offspring, an abandonment of self—a pouring forth of her most holy affections, which has been the brightest and purest portion of her history.

Love of helpless infancy—attention to its wants, its sufferings, and its unintelligible happiness, seem to form the very well-spring of a woman's heart—fertilizing, softening, and enriching all her grosser passions and appetites. It is truly an instinct in the strictest acceptation of the word. A woman, if removed from all intercourse, all knowledge of her sex and its attributes, from the very hour of her birth, would, should she herself become a mother in the wilderness, lavish as much tenderness upon her babe, cherish it as fondly, hang over it with as fervent affection, attend to its wants, sacrifice her personal comfort, with as much ardour, as much devotedness, as the most refined, fastidious, and intellectual mother, placed in the very centre of civilized society.

Instincts are those dispositions for a certain train of actions, which have been impressed, more or less distinctly, upon the minds or nature of all living beings, whether endowed with reason or not, and which, if allowed to take their own course, freed from the control of extraneous causes, are unerring guides for the attainment of particular ends. Education, change of locality and habits, will warp and derange their operations, but never produce their entire extinction. Maternal feelings and actions are rarely overcome in animals by domestication; and it would be supposed that woman, who, in addition to these instincts, which are fully as powerful within her, possesses reason, and a capability of cultivating a host of affections, would exhibit, in her social condition, maternal love in its strongest, most durable, and most amiable form.

The system of factory labour has, however, gone far towards annihilating this great and beautiful principle in woman's moral organization.

Necessity, or rather improvidence, has driven her to have recourse to labour; and that labour being continued throughout the whole day, has torn asunder those links of affection which, under almost all other circumstances, has bound a mother to her offspring; and in doing this it has deprived woman of that moral characteristic, the most influential in rendering her a loveable and loving being.

Compelled to rise early—no opportunity given for visiting home, during the day, but at the scanty and hurried meal-times—her mind and body alike enthralled by her occupation—her social affections destroyed—her frame little calculated to furnish her child with support, she becomes inaccessible to its appeals to her tenderness—leaves it to the care of a hireling, or young person, a mere infant—suffers it to be filthy and half-starved; and, as its faculties develop, takes no interest in keeping it from the contagion of vice and grossness.

A system which thus depraves and perverts, or which demands those times and occasions a mother should devote to her infant—which, from various concurring causes, so far influences her domestic habits as to interfere with the development of her social affections, and which, from its continued action upon her, at length destroys or buries them beneath a load of grossness and sensuality—must be wrong.

If a mother's love is thus injured—a love

springing as it does from the very groundwork of her moral nature, no wonder can be excited that the relations of husband and wife are perverted; and, in place of presenting a picture of what love and domestic felicity should be, developing a scene of indency—wanting everything which should render home a place for pure and chaste pleasures. The chastity of marriage is but little known or exercised; husband and wife sin equally, and an habitual indifference to sexual rights is generated, which adds one other item to assist in the destruction of domestic habits.

The delicacy of comportment, which is the palladium of married love, has no existence: separated during the whole of the day, exposed to vicious example, surrounded by a stimulant yet enervating atmosphere, the intercourse of man and wife loses its protecting influence upon the animal appetites, and ceases to become a bar to licentiousness and sexual profligacy. After their day's labour, they meet at night, not as those who love and have been separated; not to indulge in the endearments proper to their condition, but to drink and to quarrel; to behave indecently, or to taunt each other in terms of beastiality. The fountains of domestic purity being thus poisoned, the sanctity of marriage rites thus profaned—social virtue is driven from her last strong hold, and leaves them a prey to obscenity, coarseness, selfishness, and everything that is pernicious and abominable.

No amenity of manners, no gentleness of behaviour, mark the home of a factory labourer in a crowded population; no regard for conjugal obligations, no bashful reserve, no cultivation of those finer sensibilities, which can and do shed a bright gleam of pure radiance over married life, when composed of its proper elements.

If the domestic manners of the parents are thus depraved, their example cannot tend to remove the influence of the evil lessons taught their children at the mill—the gin or beer-shop—in the streets, or lodging-houses. Brother and sister lose that connexion which ought naturally and properly to exist between them: disregard for each other's welfare, a separation of interests and feelings, a forgetfulness of what is due, one to another—destroy those bonds which should link together the hearts of individuals springing from the same source, endeared as they ought to be by the memory of their younger years—years which, unfortunately for them, have been passed in total disregard for home duties, uncared for likewise by their natural guardians, and separated at an early age, to be exposed to a continuance of the same vices which deprived their homes of all beneficial influence upon their moral and social development, and inured to conduct vicious in itself and tending to destroy all the nobler and better feelings of their nature.

Neither is the conduct of parent to child, and of child to parent, a whit more engaging; but is as remote as possible from the just observance of filial and parental duties. Insubordination on the part of the child, cruelty and oppression on that of the parent; quarreling, fighting—a total alienation of affection; and, finally, a separation from home, at an age when parental control and proper domestic discipline are essential to the future wellbeing of the child.

The instances of proper feelings between parents and children are very uncommon: those offices which should be mutually rendered are neglected; an utter disregard for the well or evil-doing of either party is exhibited. If they still live together, held by mutual interests, the morals of his children form no part of the anxiety of the parent, so long as they bring their usual weekly allowance to the fund for their support. It is no unusual thing to hear a mother detail a course of wickedness and licentiousness on the part of her son, bewailing his neglected time and his lost money, without once adverting to the moral criminality of his career, or the evil of its example upon the rest of her family; or a father telling the tale of his daughter's dishonour, and cursing her for the neglect of her work, without even hinting at her depravity, or expressing sorrow for her moral declension. The brother will relate his sister's wanderings from the path of chastity without a blush; and the sister detail her brother's promiscuous intercourse amongst her own companions, without any apparent feeling of its impropriety.

A household thus constituted, in which all the decencies and moral observances of domestic life are constantly violated, reduces its inmates to a condition in nowise elevated above that of the savage. Recklessness, improvidence, unnecessary poverty, starvation, drunkenness, prostitution, filth, parental cruelty and carelessness, filial disobedience, neglect of conjugal rights, absence of maternal love, destruction of brotherly and sisterly affection, bad men, bad women, bad citizens, are its constituents, and the results of such a combination are moral degradation, ruin of domestic enjoyments, and social misery.

CHAPTER VI.

PHYSICAL CONDITION.

Connection of Food and Clothing with civilization and bodily appearance—Influence of Caste, &c., upon personal appearance—Negroes—Caffres—Bushmen—Bodily inferiority ever attendant upon barbarous habits—Personal appearance of the Domestic Manufacturer—Inferiority of personal appearance in the Factory labourer—General Ugliness of the Manufacturing Population—" Lancashire Witches" —Female Form, its Imperfections—Sexual Functions—The Pelvis—The Mammæ—Changes in these—Influence of Woman upon Man's character—Importance of Woman retaining her Social Rank—Circumstances connected with her becoming a Mother—Approximation to Barbarism—Progress of Physical declension: its probable extent—Physical Changes in the Families of the Masters.

"'WITH civilization and barbarism, food and clothing appear intimately connected. We should before-hand be induced to imagine that the most excellent development of every animated species would be effected, where all its wants were supplied, its powers all duly called forth, and all injurious and unpleasant circumstances least prevalent, and *vice versâ*. But experience teaches us that no change can by any means be brought about in an individual and transmitted to his offspring. The causes of change in a species must therefore operate, not by altering the parents, but by disposing them to produce an offspring more or less different from themselves.' Such is Mr. Hunter's view of the question, and it is certainly confirmed by every fact. Uncivilized nations, exposed to the inclemency of the weather, supported by precarious and frequently unwholesome food, and having none of the distinguishing energies of their nature called forth, are almost invariably dark-coloured and ugly, while those who enjoy the blessings of civilization, that is, good food and covering, with mental cultivation and enjoyment, acquire in the same proportion the Caucasian characteristics."

"'The different effects of the different degrees of civilization,' says Dr. Smith, 'are most conspicuous in those countries in which the laws have made the most complete and permanent divisions of rank. An immense difference exists between the nobility and peasantry of France, Spain, Italy, and Germany. It is still more conspicuous in eastern countries, where a wide difference exists between the highest and lowest ranks of society. The naires or nobles of Calicut, in the East Indies, have, with the usual ignorance and precipitancy of travellers, been pronounced a different race from the populace, because the former, elevated by their rank, and devoted only to martial studies and achievements, are distinguished by that manly beauty and elevated stature, so frequently found with the profession of arms, especially when united with nobility of descent: the latter, poor and laborious, and exposed to hardships, without the spirit or the hope to better their condition, are much more deformed and diminutive in their persons, and in their complexion much more black.' 'In France,' says Buffon, 'you may distinguish by their aspect, not only the nobility from the peasantry, but the superior orders of the nobility from the inferior, these from citizens, and those from the peasants.'' The field slaves in America,' says Dr. Smith, 'are badly clothed, fed, and lodged, and live in small tents on the plantations, remote from the example and society of their superiors. Living by themselves, they retain many of the customs and manners of their ancestors. The domestic servants, on the other hand, who are kept near the person, or employed in the family of their masters, are treated with great lenity: their work is light, they are fed and clothed like their superiors, they see their manners, adopt their habits, and insensibly receive the same ideas of elegance and beauty. The field slave is in consequence slow in changing the aspect and figure of Africa. The domestic servants have advanced far before him in acquiring the agreeable and regular features of civilized society. The former are frequently ill-shaped– they preserve in a great degree the African lip, and nose, and hair—their genius is dull, and their countenance sleepy and stupid: the latter are straight and well proportioned, their hair extended to three, four, and somesertions be doubted, when it is remembered that

of their mouth handsome, their features regular, their capacity good, and their looks animated.'"

"'The South Sea islanders, who appear to be all of one family, vary according to the degree of their cultivation. The people of Otaheite and the Society Islands are the most civilized and the most beautiful. The same superiority,' says Captain King, 'which is observable in the Eeces or nobles throughout the other islands, is found also here. Those whom we saw were without exception well-formed, whereas the lower sort, besides their general inferiority, are subject to all the variety of make and figure that is seen in the populace of other countries.*'"

Food and clothing then are intimately connected with civilization, and civilization is intimately connected with personal appearance. The half-starved and ill-clothed negro, snatched from maritime provinces, utterly barbarous in his habits, and exchanging only one sort of slavery for another, has long since stamped the notion of ugliness upon all the coloured races of men. Bruce, Denham, Clapperton, Lander, and other travellers into the interior of Africa, give assurances that there are numerous tribes, and portions of tribes, possessing in an eminent degree bodily excellence, and exhibiting in their persons but few traces of the negro lip or nose, his monkey calf or pointed shin-bone. Neither can these as-

* Elliotson—Blumenbach's Physiology.

they are described as having made some progress in the arts of life, and are distinguished into classes or castes. The Dahoman and Ashantee Caboceers, according to Bowditch and Norris, are in many points widely dissimilar from the inferior natives, —their personal appearance improving as they recede from the houseless and precarious condition of the rude African tribes.

The pastoral Caffres of Southern Africa, Mr. Barrow describes as a remarkably well-built and handsome race of men; they have made some advances in civilization, and present as perfect a picture of patriarchial existence as can well be imagined.* In all these respects they offer a striking contrast to the Bosjesman or Bushmen, who appear at some not very remote period to have derived their origin from the same source. The combined evidence of Barrow, Lichstenstein, Vaillant, Campbell, and Burchell, prove that the Bushmen are as ugly a race as any under the sun; leading a wandering life; frequently half-times even to six or eight inches, the size and shape

---

* "From the fertility of their soil, the abundance of their rivulets, and the pleasant situation of their craals, may probably arise that settled and contented turn of mind which in time of peace so remarkably distinguishes the Caffre. To him a neat little dwelling, solidly constructed in the centre of a corn-field, that was once cultivated by his ancestors, is sufficient to claim the agreeable appellation of his country. The Caffres must certainly be acknowledged a more civilized people—their industry is more perceptible, and their acquaintance with some of the necessary arts of life greatly superior."—*Le Vaillant.*

starved; then indulging ravenously, and to a most enormous extent, in eating—rivalling even the Esquimaux as to quantity; almost naked; hunted by the colonists and their dependents; their bodily form is but an index of their barbarism. There is, however, no reason to doubt but that if their wandering and predatory habits should in course of time yield to a more settled mode of life, it would be followed by a great improvement in their physical conformation.

Thus, then, it seems evident that barbarous habits, occasional want of food, bad clothing, one and all, produce bodily inferiority. The records of all nations prove this,—where an absolute division into castes is in force; by nations still termed savage, yet having amongst them a superior and inferior order, whether of governors, priests, or warriors; by slaves imported,—presenting no apparent social or bodily disparity, yet, in a few years, or in one or two generations, undergoing very decided modifications, when favourably placed for that purpose; by tribes inhabiting similar districts, and obviously tracing their origin to similar sources—the one becoming cultivators and men of settled habits, and quickly losing their coarseness of bodily form—the other leading a doubtful and precarious existence, depending for support upon the uncertain products of the chase or plunder, and retaining all their aboriginal ugliness and peculiarities.

If mankind has its origin from one common

stock—one single progenitor—from one sole Adam, according to the Mosaical and received account of creation, then must all men have the same capabilities for developing, when placed under the operation of favourable causes, personal perfection, in the understood acceptation of the word. Whether Adam was a man of colour, or whether he was a man possessing the physical traits of the division of mankind classed as the Caucasian variety, those men who have devoted the most attention to the subject, and whose opinions are of the greatest value, differ in their conclusions. If an argument might be applied, drawn from the known changes which have been observed to occur as to figure, if not as to colour, the preponderance of belief would be on the side which classed him as a genuine Ethiopian; and this, too, is the opinion which can be most completely authenticated by the present state of knowledge on these subjects. It seems likewise, on the whole, more easy to explain the conversion of a coloured race into a white, than of a white race into a coloured.

If it were necessary to bring forward as proofs, other quotations from the historians of mankind, to substantiate the position, that the manufacturing population, in consequence of their near approach to barbarous and uncivilized modes of life must exhibit great inferiority in figure and personal proportion, they might be readily found. Wherever men are condemned to toil—wherever wide distinction of caste separates them from their superiors—wherever their manners and domestic habits are coarse and improvident—their bodies uniformly become stunted, ill-shaped, wanting elegance of contour, and that development and arrangement of parts constituting beauty. The universality of this truth needs no confirmation or illustration to those who have read or inquired into the subject, or whose observation has been exercised upon a comparison of bodily forms, infinitely varied as they are, in common society. The population engaged in manufactures, since the application of the steam-engine has so assimilated and simplified their condition, offers an admirable field for examining how far external causes, aided by morals, can modify the physical proportions of man, and how far this modification is likely to influence the offspring of parents thus —changed, providing the same causes continue to operate through several successive generations.

The personal appearance of the domestic manufacturer is still well remembered by those whose memories carry them back to the days of quoit and cricket-playing—wakes—May-day revels—Christmas firesides, and a host of other *memorabilia*, now ranked but as things that were! Some faint traces of these by-gone times are still discoverable in the remoter districts: they are, however, only the "disjecta membra" of a whole division of the community.

He was a robust and well-made man, reaching

the average altitude of his race; clean-limbed, and with an arched instep; ruddy complexioned; his general contour rounded from the deposition of adipose matter, and from muscular development, possessing considerable physical power, and delighting in athletic sports. Not called into active and permanent labour till his frame was to some extent set, and his bony system perfected, he ran no danger of any shrinking or yielding in these parts; neither were his animal passions pushed into premature activity by high atmospheric temperature, aided in its exciting operation by vicious example at home and abroad. His offspring were, like himself, a chubby and rosy-faced set of urchins—healthy from their birth upward; tended by affectionate mothers, whose domestic occupations never interfered with their maternal duties. Permitted to range at will in the free and balmy air of heaven, their young minds inhaled a love of nature with their daily and hourly growth, while their social affections were fostered and directed under the eye and admonition of their parents.

The same ruddy and healthy appearance marked the whole household of the domestic manufacturer, as, at the present day, characterize the rural population of other countries enjoying moderate prosperity; they were, in fact, a strictly rural people, partly indeed engaged in the various processes of manufacture, but partly also in cultivating the soil as small farmers, either in their own right, or as leaseholders or yearly tenants of farms, averaging in extent from four to eight acres.

Their manners were, no doubt, coarse and rude, but their habits were a thousand times more moral and civilized than those of their descendants. This was owing very materially to their intercourse with their superiors, and by their living in immediate contact with them. They derived, in consequence, essential advantages from their example, which kept up a spirit of private and public decorum, quite at variance with open and avowed vice and profligacy. Hence their persons exhibited no traces of debilitating excesses; their nerves were well strung; their step elastic, and their feeling of existence buoyant and joyous. Their days were passed in sport or labour, as their necessities demanded, and their nights in deep and profound sleep, the result of active physical exertion.

The vast deterioration in personal form which has been brought about in the manufacturing population, during the last thirty years, a period not extending over one generation, is singularly impressive, and fills the mind with contemplations of a very painful character. If hardship—if low wages—if an impossibility of acquiring the means to secure the necessaries of life, or of clothing fit for protection against the influences of a variable climate—if one or all of these causes had been in operation, their effects might have been predicated. If war had swept its devastating and demoralizing wing over the land—if famine or pestilence had prevailed, the change which has been wrought might have been looked for, and its existence would have excited no surprise. But nothing of this sort has been in operation. On the contrary, with the exception of one particular class (hand-loom weavers) wages have been and are good, more than amply sufficient to supply all that is wanted even for liberal support. No 'foreign levy' or intestine commotion, save partial outbreaks of popular violence, brought about by themselves, have influenced their condition; no famine or plague has occurred to break up the bonds of society; no retrograde movement has been made to benighted ignorance, but intelligence has been becoming day after day more and more widely diffused; and yet there is a numerous population, forming a most important branch, an integral portion of the kingdom, an essential, an indispensable agent in supporting its prosperity, exhibiting in all its features an approximation to the shapes and growths of the lowest barbarism.

Any man who has stood at twelve o'clock at the single narrow door-way, which serves as the place of exit for the hands employed in the great cotton-mills, must acknowledge, that an uglier set of men and women, of boys and girls, taken them in the mass, it would be impossible to congregate in a smaller compass. Their complexion is sallow and pallid—with a peculiar flatness of feature, caused by the want of a proper quantity of adipose substance to cushion out the cheeks. Their stature low—the average height of four hundred men, measured at different times, and different places, being five feet six inches. Their limbs slender, and playing badly and ungracefully. A very general bowing of the legs. Great numbers of girls and women walking lamely or awkwardly, with raised chests and spinal flexures. Nearly all have flat feet, accompanied with a down-tread, differing very widely from the elasticity of action in the foot and ancle, attendant upon perfect formation. Hair thin and straight—many of the men having but little beard, and that in patches of a few hairs, much resembling its growth among the red men of America. A spiritless and dejected air, a sprawling and wide action of the legs, and an appearance, taken as a whole, giving the world but "little assurance of a man," or if so, "most sadly cheated of his fair proportions." Beauty of face and form are both lost in angularity, while the flesh is soft and flabby to the touch, yielding no "living rebound" beneath the finger. The hurry and anxiety of this juncture brings out very strongly all their manifold imperfections.

Lancashire has long been celebrated for the beauty of its women; "the Lancashire witches" being a standing toast in all private and public convivialities. In the higher and middle classes of society, there are certainly to be found many exquisite specimens of female loveliness—many exceedingly graceful and feminine beings. They may be seen in abundance in all the social circles, in places of amusement and parade, in which, like the sex all the world over, they naturally assemble —a passion for admiration and attention forming an essential and important portion of woman's character, and one too of the utmost value, and worthy every cultivation. But these must not be sought for amongst the precociously developed girls herding in factories. Here, on the contrary, will be found an utter absence of grace and feminine manners—a peculiar rancous or rough timbre of voice—no such thing as speaking soft and low, "that most excellent thing in woman," a peculiarity owing to various causes, a principal one of which is, too early sexual excitement, producing a state of vocal organs closely resembling that of the male. Here is no delicacy of figure, no "grace in all her steps," no "heaven within her eye," no elegance of tournure, no retiring bashfulness, no coy reserve, no indication that a woman's soul dwells there in all its young loveliness, with its host of hidden delights, waiting but the touch of some congenial spirit to awaken all its sensibilities and passions; but in their place an awkward and ungainly figure;—limbs badly moulded from imperfect nutrition—a bony frame-work, in many points widely divergent from the line of womanly beauty—a beauty founded upon utility—and a general aspect of coarseness and a vulgarity of expression quite opposed to all ideas of excellencies in the moral and physical attributes of the sex.

There is something in the female figure strongly indicative of its aptitude for the performance of certain functions peculiar to her sex. Child-bearing is one of these, and the nourishment she is subsequently destined to afford her offspring another. The gait of women who labour under any material alteration in the axis of thigh-bone is singular—a sort of waddle, an alternate side-long progression. This alteration is brought about by any change in the direction or bearing of its socket, which is seated in the pelvis—or by changes operated upon the thigh-bone itself—or by changes in the direction of the action of certain great muscles which have their origin on the spine, pelvis, or adjacent parts,—all of which are liable to be influenced more or less by the circumstances in which she spends her infant years, and by that derangement in the prima viæ, and their effects, which her occupation and habits expose her to. This gait may be detected in great numbers of factory girls and women, and is exceedingly ungraceful—ungraceful in itself, and still more so in its impression upon the mind, by the evidence it gives of certain alterations in form peculiarly unsexual. Neither is the condition of those organs from which the child is to derive its first aliment less strikingly illustrative of their habits. Very early in life, from ten to fourteen years, the breasts are often found large and firm, and highly sensitive, whilst at a later period—at a period indeed when they should shew the greatest activity and vital energy—when in fact they have children to support from them, they are soft, flaccid, pendulous, and very unirritable —both states giving the most decisive proofs of perversion in the usual functional adaptation of of parts.

The moral influence of woman upon man's character and domestic happiness, is mainly attributable to her natural and instinctive habits. Her love, her tenderness, her affectionate solicitude for his comfort and enjoyment, her devotedness, her unwearying care, her maternal fondnesses, her conjugal attractions, exercise a most ennobling impression upon his nature, and do more towards making him a good husband, a good father, and an useful citizen, than all the dogmas of political economy. But the factory woman cannot have this beneficial agency upon man's character. Her instincts, from their earliest birth, have been thwarted and pushed aside from their proper channels; they have had no field in which they could be cultivated, no home where their aberrations might have been checked, no legitimate object on which her love could be lavished; on the reverse,—her passions have been prematurely developed, her physical organization stimulated into precocious activity, her social affections utterly blighted, her person rendered uninviting by its want of feminine gracefulness; her occupation has destroyed her home sympathies and maternal affections, and finally, in place of seeking her pleasures, enjoyments, and happiness in ministering to the wants and welfare of her household, she seeks her gratification in other, less pure and less womanly stimuli, fatal alike to her health, comfort, and appearance.

Under these circumstances, woman is reduced to precisely the same grade in the social rank which she holds in half civilized countries. She is no longer the companion of man, in the proper meaning of the term; but is a mere instrument of labour, and a creature for satisfying his grosser appetites. It is true, her condition differs in some respects from that of her sex in countries which are called savage; she is not the slave, though she is no longer the companion of man; neither is she the subordinate being whose interests are esteemed so secondary as to be totally unregarded;—but, as far as concerns the better portion of her attributes, she is upon the same degraded level.

Nothing would tend more to elevate the moral condition of the manufacturing population, than the restoration of woman to her proper social rank· nothing would exercise greater influence upon the form and growth of her offspring, than her devotion to those womanly occupations which would render her a denizen of home. No great step can be made till she is snatched from unremitting toil, and made what Nature meant she should be—the centre of a system of social delights. Domestic avocations are those which are her peculiar lot. The poor man who suffers his wife to work, separated from him and from home, is a bad calculator. It destroys domestic economy, without which no earnings are sufficient to render him comfortable; it produces separate interests, and separate sets of feelings—they lose their mutual dependence upon each other—their offspring is suffered to starve or perish—to become, even as a child, the imitator of their bad example—to have its frame permanently injured—to acquire bodily conditions which it must, in its turn, transmit to its own children—for, by their being thus early implanted, they become part and parcel of its very nature—till, in the end, a point of physical declension will be reached, from which the return to a condition approximating to that of the more perfect and valuable part of the species, must be by slow and painful gradations.

Another circumstance, which shews very decidedly the extent of moral and social degradation in the female sex among this population, and which reduces them below the level of the untutored savage, is this—that upwards of two-thirds of all the children born to this class in Manchester are brought into the world by the aid of public charity; that decent and proper pride, which should lead women to prepare for an event so interesting to her as a mother, is abolished; and, like the Indian Squaw, she pursues her labour almost till the hour of her delivery—to abandon her tender and delicate infant, after the interval of a few days, to other and hireling hands;—again to pursue her usual routine of work.*

* "The number of married women delivered annually by the Manchester Lying-in Charity on the average of the last two years is upwards of four thousand three hundred. Now it is well known to those conversant with statistics, that the number of baptisms occurring annually in England is in the proportion of one baptism to thirty-four inhabitants nearly. This is about the average for the kingdom. But the births are somewhat more numerous than the baptisms, because some children are still-born, and others die before they are baptized, and some are christened who are not entered into the public registers. When we have made allowances for these incidental circumstances, the annual rate of births in Manchester may be stated to be, at the very highest, as one to twenty-eight of the existing number of inhabitants—perhaps one to thirty would be nearer the truth.

"The application of this, with the view of showing what proportion of the inhabitants produce the number of births, attended annually by the Lying-in Charity, is a sufficiently simple process. Taking it for granted that the four thousand three hundred births before-mentioned as the average annual number that may be anticipated regularly to occur (and I am sorry to say that there is no prospect of the number decreasing),† it follows that since the production of one birth annually requires twenty-eight inhabitants—the production of four thousand three hundred births will require one hundred and twenty-four thousand four hundred inhabitants; in other words, if we take the population of Manchester at two hundred and twenty-seven thousand, we have thus considerably more than half of

† Mr. Roberton's anticipations have proved correct enough, the present number of women delivered by the charity being very near 5,000.

The squalid and mean appearance already acquired by the inferior order of manufacturers—even though one generation has scarcely passed away since the processes in which it is engaged have been so essentially changed, by the application of steam—is not the less remarkable for its own sake, than as it arouses examination into the probable alterations their descendants must undergo, placed, as they inevitably must be, as still more abject and passive slaves beneath machinery.

Physiological inquiries will serve to develop these changes to some extent; facts of observation are likewise in abundance,* and both prove, that a body worn down and debilitated, although the generative faculty may be uninjured as to intensity in either sex,—cannot give the necessary pabulum for the production of a vigorous offspring, endowed with active vitality. Bodily deformity, bodily defect, or ugliness, may not, nay, in general, will not be transmitted. But an universal weakness and want of tone in all the organs, a disposition little able to resist disease, sufficiently

---

the whole, who are in so destitute a condition, or, if you will, so degraded, as to have their offspring brought into the world by the aid of a public charity."—*Roberton on the Health*, &c. p. 22.

* The lines of Horace are founded upon nature and her workings—

> " Fortes creantur, fortibus et bonis;
> Est in juvencis, est in equis patrum
> Virtus."

prove, that although the child has not inherited the peculiar failings of its parents,—yet it has, as an heir-loom, their weakened constitution, attended with all its liabilities to physical inferiority. That parents so circumstanced, may occasionally produce fine and robust offspring, goes for nothing; the mass must be depraved and deteriorated, and how far this may extend, no human sagacity can foretel.

Nature may, indeed, raise up some countervailing influence ; her powers of modification and adaptation are extraordinary, and their extent unknown ; but neither the extent nor effect of this can be predicated. Judging from those things, upon which judgment should alone be formed, namely, an accurate examination into the existing state of things, and the causes which have led to that state, and proceeding upon the belief that similar causes must be attended with similar results, it may be asserted,—that a diminutive and ill-shaped race, possessing little muscular strength, and fitted as labourers *only* to act in subservience to a more powerful agent, will be the product of the present mode of manufacture. Steam, which has been the principal instrument in rendering human force valueless, will, in time, be the means of generating a set of helpers, peculiarly its children, and doomed to depend entirely upon it for support.

The want of sympathy and proper intercourse between the master manufacturer and those employed by him, is one very powerful cause which tends to keep down their moral habits and their physical condition. Living, as the majority of the operatives do, in the midst of large towns, and incessantly occupied, they see nothing around them but beings like themselves. There is no middle class to which they can look for countenance and example. The family of their employer they either detest, and studiously avoid imitating its manners, however humbly or remotely, or they see so little of it, and are treated so coldly and harshly by it, that they nurse their own grossness in direct and avowed opposition to it.

The force of external circumstances in modifying bodily form, is seen equally forcibly, in another and more pleasing point of view, in the manufacturing districts. Many of the masters have raised themselves from the very humblest rank of labourers—in many instances after a family had been born to them in their humility. These individuals with their families, at this period, of course, possessed all the traits distinguishing their grade, both moral and physical. Change of condition, better food, better clothing, better housing, constant cleanliness, mental cultivation, the force of example in the higher order of society in which they are now placed, have gradually converted them into respectable and even handsome families. The first remove places them still more favourably, and *ceteris paribus*, they become elegant and intelligent females, and wēll formed and robust men. They now resemble but slightly, in their general aspect and deportment, the class from which they have risen ; almost as slightly, indeed, as the pupa or chrysalis resembles the gaudy and summer painted-butterfly or moth which springs from it.

CHAPTER VII.

INFANT LABOUR.

Infant Labour—Small and Simple Machinery—Mill Labour by Children from Five to Twelve Years of Age—Apprentice System—Size and Locality of the First Spinning Mills—Excess of Labour—Dr. Percival's Inquiries in 1796—Sir Robert Peel's Evidence—Free Labour—Introduction of Steam power—Enlargement of Machines—Great improvement of Mills—Proportion of young Children employed in Factories—Increase of adult labour—Effects of improved Machinery upon adult labour—Cause of Women and Children being preferred by Masters—Equalization of Infant and Adult labour—Material organization of Childhood described—Progress of Ossification—State in the Embryo, in the Fœtus, in the new-born child, described—Periods of completion of different points according to function—Progress of Ossification subsequent to Birth—Physical conditions, &c.—Childhood—Importance of proper Nutrition—Necessity for Exercise, &c. &c.—Condition of the Factory Child—Effects of disturbing healthy Nutrition—Earthy deposit in Bone—Rickets, Scrofula, is hereditary, in what sense—State of Factory Child, prior to commencing Labour—Necessity of this being properly understood—Error of Abolitionists—Healthy children fit for labour at eight or nine years of age—State of a Child from the country—Its change when placed in a large town—Difference of capability for bearing labour in a Town and Country Child at nine years of age—Their Conditions compared—Factory Labour, consequences to Children described—Deformity, how produced—Inhalation of dust, Effluvia of oil, &c.—Odours influence upon Health—Temperature—First impression upon Children—Ventilation—Deoxigenation of respired Air—Continuance of Mill Labour—Consequences—Examination of Children—Opinion of Dr. Baillie—Influence of Deformity on the Female—Mischiefs of present social condition—Difficulty of alteration—Want of restriction, &c. &c.

"The discoveries of Arkwright, Watt, Crompton, and other great benefactors of mankind," remarks an intelligent and extensive manufacturer, in his evidence before a Committee of the House of Commons, on the employment of Children in Manufactories, "produced a complete revolution in the spinning manufactures of cotton, of wool, and of flax, which might now be more properly termed mechanical productions: by the application of the power of waterfalls, and of ingenious mechanism, adults were superseded by children, whose wages were lower, and who soon acquired a great dexterity: the production of the manufacturers was enormously increased, because the extraordinary cheapness produced a consumption in proportion to the great augmentation of the wealth and revenue of the nation. The very moderate exertion of the children, and the great expense of the mechanism, introduced the custom of working twelve hours per day in the mills. Under these circumstances, it became the duty and the advantage of every proprietor, to render his mill as healthy as possible, by cleanliness, ventilation, spaciousness, and temperature, and to interest himself generally for those employed by him. As the children grew to be adults, notwithstanding their acquired expertness, they became too big for the machinery, and their labour too expensive. In consequence of the introduction of the revolving steam engine, factories were established in manufacturing towns: the establishments in manufacturing towns were more calculated for adults, and machines of greater size and complication were constructed for the more difficult and finer kinds of spinning. If this arrangement had not taken place, and the business continued to extend, a numerous population of children would have been formed, and employed in remote situations, without an adequate demand for their labour when adults;—the employment of adults by such suitable establishments in manufacturing towns therefore obviated this evil. It has likewise produced that excellence in the operations which could never have been obtained or rewarded, except in the great and open market of talents, in large towns. The  labour of adults is infinitely more regular in machinery than in manual employment, *but can never be controlled and reduced to the uniformity of attendance of children.* On the other side, they are capable of greater exertion, and possess far greater skill. The system of employing adults, restores society to its proper direction, which had been changed by sudden discoveries, and probably neither a younger nor a greater number of children will be required to be trained in the mills upon the new scale, than that proportion which is necessary to supply the demand for adults in many other labouring classes of society."

The simplicity of the first machines adapted for spinning, and their small size, fitted them for being tended by children. The localities in which the mills were at first chiefly erected, for the convenience of water power, were often remote from towns or villages, from which alone an adequate supply of hands could be obtained: hence, in the early period of mill labour, apprentices, from six to twelve years of age, were almost the only workers. These apprentices were chiefly taken from the workhouses of large towns, such as London, Birmingham, &c., and from foundling hospitals, and transmitted in droves* to the different mills, where, in many instances, it is to be feared, they suffered very severely. Villages, however, sprung up in the vicinity of the mills, and the parents of children very naturally took advantage of their labour, and hence, free labourers became superadded, and, in time, displaced the apprentice system, to a considerable degree. The construction of the first mills was, of course, fitted only for small machines; they were consequently small, and the rooms in them were low, and of very contracted dimensions, and very little precaution used, either as to ventilation or temperature. The time of labour was extended to twelve hours, with very little interval; the immense profits which accrued from their produce, pushing aside all ulterior considerations. Nor was this all: unsatisfied with the day labour, the night was almost uniformly spent by one portion of the hands in the mill; the owners or occupiers thus

* Mr. Arkwright's first mills were almost entirely filled with very young children. The house of Peel and Co. employed at one time a thousand of these children.

securing twenty-three hours out of the twenty-four, for making his machinery valuable.

There cannot be a question but, that child-labour, urged to this extent, and under these circumstances, was prejudicial in every way; and in 1796, the evil of it had aroused the attention of various philanthropic individuals, amongst whom the name of Dr. Percival stands conspicuously forward, and he, in conjunction with others, stirred themselves to produce some alteration and amendment. On the establishment of the Manchester Board of Health, the following statement was made by him, on the 15th of January, 1796: "The Board have had their attention particularly directed to the large cotton factories, established in the neighbourhood and town of Manchester; and they feel it a duty incumbent on them, to lay before the public the result of their inquiries:— 1. It appears that the children and others, who work in the large cotton factories, are peculiarly disposed to be affected by the contagion of fever; and that when such infection is received, it is rapidly propagated, not only amongst those who are crowded together in the same apartments, but in the families and neighbourhoods to which they belong. 2. The large factories are generally injurious to those employed in them, even when no particular diseases prevail, from the close confinement which is enjoined, and from the debilitating effects of hot or impure air, and from the want of the active exercises which nature points out as essential in childhood and youth, to invigorate the system, and to fit our species for the employments and the duties of manhood. 3. The untimely labour of the night, and the protracted labour of the day, with respect to children, not only tends to diminish future expectations, as to the general sum of life and industry, by impairing the strength, and destroying the vital stamina of the rising generation, but it too often gives encouragement to idleness, extravagance, and profligacy in the parents, who contrary to the law of nature, subsist by the oppression of their children. 4. It appears that the children employed in factories, are generally debarred from all opportunities of education, and from moral and religious instruction."

According to the evidence of Sir R. Peel, who was himself very extensively engaged in manufactures at this period, contagious and epidemic diseases were very common in the people, or rather children, employed in these badly constructed mills, and under the influence of such continuous labour. He remarks that they were uniformly of stinted growth, and that although in his own case they were fed and lodged under his roof and at his own expense, they still looked to be in ill-health. That gross mismanagement existed in numberless instances there can be no doubt—and that these unprotected creatures, thus thrown entirely into the power of the manufacturer, were over-worked, often badly fed, and worse treated. No wonder can be felt, that these glaring mischiefs attracted observation, and finally led to the passing of the Apprentice Bill* —a bill intended to regulate these matters. It may be questioned whether the bill of itself did much good, but from the discussions it excited, and the vast superiority displayed by some mills over others, when general inquiries were made,—very beneficial consequences were the result, and some attention began to be paid both to the better ordering of the mills, and to the welfare of the children. As these establishments went on increasing, a new population sprung up around them, and in a few years free labour quite displaced the prior hands. The machines for spinning were day after day becoming more bulky, and requiring greater skill and exertion for producing fine numbers, so that adults gradually found their way to them. The application of steam as a moving power, which became general from 1801 to 1804, produced a great change in all respects. It did away with the necessity for so much water power, and hence mills were more commonly built in towns or populous districts favourable as to coal, &c., and where a population was at hand for their occupation. The uniformity, the increased rapidity of motion, and the greater size of the machines, called in some departments for active labour, and grown up men and women

* The Apprentice Bill was passed 1802.

were now largely engaged in spinning;—but still children formed the majority of hands, many processes being better calculated for them than for adults. But they were not put to work quite so young, few before ten years of age—which in fact the masters found to be as soon as they could be properly useful.

The hours of labour have not undergone any very material alteration, since the universal application of steam, and since the doing away with night labour as a general custom. Notwithstanding this the condition of the children has undergone great amelioration. This has arisen not from increased wages, but from the improvement in the construction of the mills. Subsequent to 1806, when the steam loom was first brought into operation, many of the first mills were either much enlarged, or in very numerous examples abandoned by the more wealthy manufacturers, and in their stead large buildings were erected, fitted to receive, in addition to the spinning processes, a quantity of looms. Nor was this increase in size the only advantage. The rooms were much more lofty, generally of large proportions, had numerous windows, so arranged as to afford excellent ventilation, whilst improved modes of warming them did away with many of the inconveniences which had hitherto attended upon this necessary portion of their interior economy. This consisted in carrying through the rooms pipes conveying steam, or heated air, which enabled the overlooker to regulate the temperature according to the necessity of the season. Various improvements took place in the cleaning the cotton, so as to free the mill from a great part of the flue or fine cottony down, which in the old mills, spinning coarse numbers, had been nearly suffocating: cleanliness began to be cultivated, and this was materially aided by the use of cast iron floors, which were much more easily cleaned than wood, which had absorbed the oil droppings, and soon became offensive and exceedingly difficult to get off; added to which was the gradual substitution of iron in the framework of machinery, which by its diminished bulk materially increased the empty space in the rooms, the wooden frames having been large and massy, and always getting filled with dirt.

The number of young persons employed in factories since the breaking up of apprentice system, and since the divisions of labourers have fairly settled themselves, has not till of late years varied very largely, when estimated as to the proportion they bear to the remainder of the hands. In 1816, a return from forty-one mills, in Scotland, gave a total of 10,000 persons engaged in them. Of this number

Under ten years of age.	From ten to eighteen years of age.	Above eighteen years of age.
191 males. 224 females.	1,179 males. 2,810 females.	5,596
415	3,989	
Total under 18, 44,404		

A similar return from 48 mills in Manchester gave a total of 12,940 hands.

Under 10, 793; 10 to 18, 5460; above 18, 6687.

This approximates very nearly to the table first given.

A return from 36 mills in Stockport, in 1833, gave a total of 11,444.

Under 10, 320; 10 to 18, 4017; above 18, 7,101.

The diversity in the foregoing tables arises from several causes. Wherever spinning is more especially prosecuted, there the number of hands under eighteen years of age will appear porportionately increased; and wherever weaving is the staple manufacture, there the number of hands upwards of eighteen will be in greater ratio. There can be no question indeed, but that since steam-weaving became so general as to supersede the hand-loom, the number of adults engaged in the mills has been progressively advancing; inasmuch that very young children are not competent to take charge of a steam-loom. The individuals employed at them are chiefly girls and young women from sixteen to twenty-two or twenty-three years of age; indeed the weavers in many mills are exclusively females, and it is not likely that any very material alteration in this respect will occur. In spinning, children are as valuable workers in many of its processes as adults, in some degree more so.

Yet a change is gradually creeping over the condition of the operatives, and a disposition is developing itself to have recourse to the labour of women and children in preference to adults. The causes which have led to this are, the great improvements which are taking place in machinery, and its application to an infinite variety of minute operations requiring the nicest management, and the requisite power being given by steam.

The labourer is indeed become a subsidiary to this power. Already he is condemned hour after hour, day after day, to watch and minister to its operations,—to become himself as much a part of its mechanism as its cranks and cog-wheels,—already to feel that he is but a portion of a mighty machine, every improved application of which, every addition to its Briareus-like arms, rapidly lessen his importance, and tend to drive him from a participation with it, as the most expensive and unmanageable part of its materials.

The contest however between human and steam power, as applied to manufactures, is not yet so far advanced as to annihilate its value to its possessor. In another portion of this work it is satisfactorily shewn that the labourer, though losing rapidly his independent character, is paid for his labour a sum amply sufficient, when properly applied, to supply all his natural wants, and to provide him with comforts and opportunities for making provision for sickness or old age, without becoming a burden upon the fixed capital of the nation. This observation, it must be understood, refers to that portion only of manufacturers engaged in power-loom weaving and spinning with their various processes. The hand-loom weavers are, on the contrary, crushed by their mighty opponent to the dust, and keep up a warfare with it upon the most unequal and untenable grounds.

The necessity for human power thus gradually yielding before another and more subservient one, has had, in the first place, the effect of rendering adult labour of no greater value than that of the infant, or girl; mere watchers, suppliers of the wants of machinery, requiring in the great majority of its operations no physical or intellectual exertion; and the adult male has begun slowly to give way, and his place been supplied by those who in the usual order of things were dependent upon him for their support.

It may be asked why, if adult labour was still efficacious, though its physical and intellectual energies were not demanded, it was thrown aside, and women and children brought to perform its functions; thus reversing the law generally acknowledged, that man, as a husband and a father, should furnish his wife and family with support, till such periods as his offspring were themselves arrived at an age, when the development of their physical powers fitted them for earning a separate and independent support, or at least assisting their parent, and lessening in some degree his burdens. Why take from their homes and proper occupation women and children, and employ them to the exclusion of their husbands and fathers, and subject them to labour which, in the instance of the children, their physical organization was unable to bear, and in that of the women must of necessity lead to the neglect of all the domestic offices—household economy—and conjugal and maternal duties; why throw the active labourer into idleness and dependence, and thus lower immensely his moral importance? and why thus, laying the axe to the root of the social confederacy, pave the way for breaking up the bonds which hold society together, and which are the basis of national and domestic happiness and virtue?

The previous sketch of the progress of the ma factures in reference to machinery and its influence upon labour, and the conditions and localities in which the mills were first established, will explain the employment of children in the spinning department to the exclusion of adults; the introduction of the steam-loom, the subsequent excellence and simplification of the machinery—and lastly, a history of the combinations, will be sufficient to unravel the question, and to explain why masters, taking advantage of steam power, and finding that the child* or woman was a more obedient servant to himself, and an equally efficient

* "The fingers of children at an early age are very supple, and they are more easily led into the habits of performing the duties of their station."—*Evidence before Committee.*

slave to his machinery—was disposed to displace the male adult labourer, and throw him for support upon his family, or send him, as it might happen, a pauper or depraved being to prey upon the community at large.

It may be supposed that the masters, in thus wishing to rid themselves of a turbulent set of workmen, were influenced by other causes—that women and children could be made to work at a cheaper rate, and thus add to his profits. But this is not the case. Out of 800 weavers employed in one establishment, and which was then (four years ago) composed, indiscriminately, of men, women, and children—the one whose earnings were the most considerable, was a girl of sixteen years of age—a stunted, pale, and unhealthy looking creature, apparently totally unfit for work of any sort. This arises from the mode of payment now universally adopted by the trade, and which is payment for work done--piece work as it is called; the spinner for the number of pounds of yarn he produces; the weaver for the number of cuts or yards turned off from his loom. Thus the active child is put upon more than a par with the most robust adult;* is, in fact, placed in a situation decidedly advantageous compared to him. It is true there are some of the processes which do require the exertions of strength and skill, and in these, of course, men are still em-

* "The work is full as well done by children: is better done by children."—*Evidence before Committee.*

ployed; but these processes are but few, and even now machinery is making them still fewer.

The master, therefore, whose hands consist principally of children and women, produces as much cloth or twist, and pays for it the same amount of wages, as if his hands were exclusively male adults.

An objection which is frequently made as to the salubrity of factory labour, based upon the comparatively very small number of men upwards of forty engaged in it, receives elucidation from this circumstance:—Workmen above a certain age are difficult to manage. The overlookers, in most mills, are young men, holding their situation by their activity and strictness of inspection. Men who come late into the trade, learn much more slowly than children; and even when capable of managing and understanding all its details, turn off a much less proportion of work than mere infants; and as all are paid alike, so much per lb. or yard, it follows that these men, by producing much less, are not more efficient labourers than girls and boys, and much less manageable.

It has happened, then—that in consequence of the great improvements in the adaptation of machinery to complicated and delicate processes, and to having, in the steam-engine, a moving power capable of exercising any requisite degree of force,—that machines, thus impelled, requiring merely feeders or watchers; that these watchers or feeders, being not, of necessity, strong in body or intellect; that adult male labour having been found difficult to manage, and not more productive,—its place has, in a great measure, been supplied by children and women; and hence the outcry which has been raised with regard to infant labour, in its moral and physical bearings. The moral considerations have been partly discussed in another portion of this work.* The important question of infant labour, as to health and physical development, two points intimately connected with the future welfare, well-being, and happiness of the individuals, now remains to be examined.

The material organization of childhood is decidedly unfavourable to labour. The functions of nutrition, which are those peculiarly active, leave but little of either nervous or muscular energy, beyond what is required for their due supply. The whole body is in a state of rapid alteration; full of vascular excitement, and requiring long periods of repose. The constitution at large is very excitable, abounding in vitality, and prone to irregular action. The osseous system is incomplete; its structure, as yet, being in a very great proportion cartilaginous, and hence beautifully adapted for accommodating itself to the growth and extension of the body. It is, however, soft, yielding, bends beneath pressure, and is easily made to assume curvatures and alterations in direction, incompatible, more or less, with

* Vide Chapters III. IV. V.

the natural arrangements, which are, in all instances, full of harmony and architectural and mechanical beauties. The long bones, those of the arms and legs, do not become completely ossified till the body has attained its full growth. The flat bones—those of the head, pelvis, &c., become hard sooner, serving, as they do, for the walls of cavities or supports for delicate and important viscera, essential to life.

It will be useful to trace the progress of bone from its mucous state in the fœtus to its complete perfection in after-life, as it bears very considerably upon the fitness or unfitness of young children for labour—and will sufficiently demonstrate that mechanical causes may have a very important influence upon its due development.

The first stage in which the anatomist can detect bone, or rather the rudimentary parts which are subsequently to become the depositories for bony matter, is at a very early period—at a time when the whole embryo is a mere mass of transparent matter, which is so soft and unresisting, that it yields beneath the finger like a firm pulp. At this time, however, the forms of the bones are discernible, holding their proper respective positions in the frame of the fœtus. This is the condition in which they will be found in the very early periods of pregnancy. Gradually they become firmer—more cartilaginous—and about the end of the first month of utero gestation, specks of bony deposit appear in the centre of the long bones, and in one or two or more places in the flat bones. From these specks, as from centres, radiate osseous fibres, towards the extremities of the various separate portions, into which the long bones and the flat bones, forming the cavities, are divided. In the long bones these osseous striæ, or streaks, are longitudinal; running towards their extremities ;—in the flat bones, they take a star-like course, diverging towards their edges ; and in the short bones, such as the bones of the hand and foot, they proceed from the interior to the surface. Previous to the deposit of bone, the cartilaginous parts were nearly without colour, just indeed tinged with a yellowish hue ; but now a red dot is perceived in the point where the bone is forming, and blood-vessels gradually shew themselves. This red spot is synchronous with the earliest bony formation ; not that the cartilaginous parts were before destitute of circulating fluids, but that its circulation consisted of colourless matter. This bony formation steadily progresses till the period of birth, and has generally touched, in one or more points, the extreme length of the long bones, and the edges of the flat and small bones. But the whole is very far from being complete. The heads or extremities of the long bones offer distinct points of ossification ; in the flat bones—those of the head and pelvis, and shoulder blades, several distinct points are also often found ; each forming a small centre for the divergence of similar striæ to those of the primary ones. The bones of the head, which in later life become one continuous case, or surface, are at this period divided into several plates, that gradually stretch out their ossific rays, and indent into the edges of those next them, forming serræ, or teeth, which dovetail one into another, and finally make a very firm and immovable union, the bone breaking before these give way, as the edges are thickened purposely to strengthen and fortify this point of junction.

The Divine Architect, whose omniscience has contrived a mechanical apparatus like the human body, has also, in the mode of its growth, and the periods appointed for the completion of particular parts, clearly indicated the order in which their functions should be called into active employment. Respiration being immediately essential to the new-born creature, the points of bones to which the muscles are attached, chiefly concerned in that office, are the very first to shew signs of ossification. So early as the end of the first month of gestation, the collar-bones and ribs are hardening, and at birth are sufficiently firm to afford unyielding fulcra for them to act upon. The spinal column shews early marks of ossification, as it is in a considerable degree interested in the same function; while the extremities of the long bones and the edges of the flat bones are still incomplete, and the process of hardening in the small bones of the hands and feet is hardly commenced.

At this period (the time of birth), the osseous

system is a great series of detached portions of bone, incapable of supporting the soft parts without yielding beneath their weight—the connecting medium between them (cartilage) is flexible; the only portions which have assumed a very decided formation being those connected with respiration.

Month after month, as the infant acquires muscular vigour, the bones solidify in ratio as the peculiar organs are called into play. Point after point hardens, and is connected with the shafts of the long bones, or firm ridges spring up along the surface of the flat bones, giving origin or insertion to the moving powers which are destined to enable the child to perform the actions of locomotion, lifting weights, &c. &c.

It is not till complete growth has been attained, which is from the seventeenth to the twentieth year, that all the bones are completely ossified, and till this period they are more or less liable to give way under continued pressure. They have, however, assumed a very considerable firmness of structure prior to this, and little danger need be apprehended of evil consequences from any common labour or exertion which the purposes of life require.

The system of bones—the skeleton—is the basis or frame-work on which all the soft parts are, as it were, hung. It sustains their weight, affords points of support for muscular action, and by the beauty and order of its articulations, is capable of assuming a variety of directions, while its symmetrical proportion is most perfect, and admirably adapted for the packing and convenient arrangement of all the numberless components of which a man's body is made up.

During the growth of the child, then, its bones are to some extent soft and yielding,—a circumstance familiar to many mothers, in whose infants the bony structure has not, from some fault in its peculiar nutrition, gone on apace with muscular development; and which, by inducing the child to support its weight upon its feet, its legs give way beneath it and bow out,—a very common case in strumous families.

It has been said that the child's system was one full of excitement, of rapid changes in structure, and that its sensibilities were acute. The world brings to the infant, for many years, a constant succession of new sensations, and its activity, its variableness, its jocund laugh, its quiet repose, are all evidences that, like the young of other animals, its life was intended by the Creator for unrestricted bodily enjoyment, for to it motion itself ever seems a high gratification.

No one familiar with the wants and habits of a child is ignorant how irksome confinement is to it—how utterly unfit it is for settled occupations —how listless and heartless it speedily becomes when shut out from its instinctive mode of life—how soon it droops and pines, and loses the vigour and freshness of the wild flower.

Whatever causes tend to derange digestion must

of necessity interfere with nutrition; for in fact the terms may be considered as almost synonymous. The growth of every part depends intimately upon this first process being well performed. Whatever interferes with this, deranges the whole circle of the animal economy, and if continued, will lay the foundation for a series of irregular actions, which may either run into scrofula, if such taint already exists, or may bring about, independently of such taint, the same or some analogous affection. Whatever vitiates this series of operations—and it is liable to be vitiated by many causes—puts the whole animal machine out of equilibrium; and irregular formations, strumous swellings of the concellated heads of the long bones, as at the knees, ancles, elbows and wrists, or a similar train of evils in the glandular structure, evidenced by swellings in the neck, terminating in sores difficult to heal, or of the mesenteric or other glands of the abdomen, or of derangements in the mucous surfaces of the mouth, eyes, breathing apparatus, bowels, &c. &c.—all these are results, more or less marked, of any continued derangement in that most important function of childhood—nutrition.

Light, air, recreation, freedom from toil, proper clothing, wholesome food, attention to the bowels, &c., are all essential to this function being well performed. Children placed even under the most favourable conditions, if deprived of their necessary amount of unrestricted physical enjoyment, shew very plainly their want of it. In fashionable boarding-schools, and in private families, where tuition is commenced early, and where little proper exercise is taken, but where every attention is paid to food, &c., the pallid face and spiritless demeanour of the children very soon shew that nothing can compensate for the abstraction of their necessary stimuli.

The factory child,* from the very earliest period of its existence, is, in large towns subjected to all the necessary causes for the production of the physical evils, resulting from derangement of the digestive organs. Its food is coarse, and the times of feeding irregular; it is exposed to cold, ill clad, allowed to be filthy. It inhales the impure atmosphere of a badly ventilated and uncleanly house, and the equally impure one of an unpaved and unsoughed street; is allowed to get wet, goes barefoot, no attention is paid to its dietetic comforts, and consequently vast numbers die very young, and the remainder who live exhibit symptoms that they have suffered from neglect and exposure. They have pale and flaccid features, a stunted growth, very often tumid bellies, tender eyes, and other marks that the *primæ viæ* have been permitted to go wrong. The effect of all the other causes is aided by the abominable practice, so general amongst this class, of cramming their

* Vast numbers of these children are hired out at the rate of 1s. 6d. per week.

children with quack and patent medicines, to quiet their irritability, increased as it is by their digestive derangement, or the equally pernicious one of giving them gin for the same purpose.*

Whatever tends to disturb that function, the products of which furnish the materials for the building up of the body, or which alters, and depraves those materials, must necessarily im-

---

* "In consequence of the absence of the mothers, these children are entrusted, in a vast majority of cases, to the care of others, often of elderly females. These women often undertake the care of several infants at the same time; the children are restless and irritable from being deprived of a supply of their natural food (as when the mothers suckle them they can only perform that duty in the intervals of labour), and the almost universal practice amongst them is to still the cries of the infant by administering opiates, which are sold for the purpose under several well known and popular forms. The quantity of opium which from habit some children become capable of taking is almost incredible, and the effects are correspondingly destructive. Even when the children have a healthy appearance at birth, they almost uniformly become in a few months puny and sickly in their aspect, and many fall victims to bronchitis, hydrocephalus, and other diseases, produced by want of care, and the pernicious habits we have detailed. We may mention, also, that spirits, particularly gin, are given frequently when the infants appear to suffer from pain in the bowels, which from injurious diet is very common among them."—*Inquiry, &c.* p. 17.

"The child is ill-fed, dirty, ill-clothed—exposed to cold and neglect, and in consequence more than one half of the offspring of the poor die before they have completed their fifth year. The strongest survive, but the same causes which destroy the weakest, impair the vigour of the more robust, and hence the children of our manufacturing population are proverbially pale and sallow, though not generally emaciated, nor subjects of disease."—Dr. Kay's pamphlet, p. 70.

pede and retard the completion of the healthy structure of the system. In childhood, much has to be done in these respects; muscles have to be strengthened and brought into use; the bony fabric solidified; growth carried on, by the deposition of new matter in every portion of the body; preparation made for sexual development: all these keep the feeders in incessant action. The vessels which are to be the first stage in preparing the new matter, and carrying it to its final depositories, are all connected with the alimentary canal: these convey the chyle they have elminated from the food. If, however, this chyle is imperfectly formed, which must inevitably be the case when digestion is impaired by the causes already mentioned, it cannot fulfil perfectly its destined offices in the system—growth is retarded, and particular structures unfinished.

The formation of bone,—the necessary earthy deposit upon which its hardness depends, is one singularly under the influence of the digestive operations. Ricketty children are ever wrong in this respect: the food they take furnishes but little nutriment; their bowels are disturbed; assimilation goes on imperfectly; all the processes connected with absorption and deposition are irregular, and do not preserve that reciprocity of action, that balance of quantity, which is necessary for proper growth.

Without going more diffusely into the pathology of the two diseases, scrofula and rickets, which are in their origin clearly traceable to disturbed digestion, it may be sufficient to state, that both, when they have been fairly established, seem to be transmissible from parent to child—to become hereditary. By the term hereditary, however, when applied to diseases, whether mental or physical,—it must be understood only in a general sense,—as meaning, that a parent endows his offspring with certain peculiarities of constitution, certain tendencies to take on diseased actions—not that he transmits the disease in its distinct form, but that he gives a liability to have it called into action by causes which would not develope a similar disease in a child descended from parents perfectly free from any such taint. With scrofula this is decidedly the case—scrofulous parents very frequently having offspring with all the traces of its peculiar diathesis—light eyes, fair skin, &c. &c. With rickets this transmission is less observable, probably because its incipient stages are not so strongly marked.

The factory child then, from its birth, is placed under circumstances certain to deprave its nutritive functions, and to produce effects injurious to the healthy stamina of its physical organization. Its soft parts become flabby, relaxed, open to disease; its osseous system becomes firm very slowly and irregularly, yields to superincumbent weight, producing deformity more or less marked, and more or less prejudicial, as affecting parts more or less connected with important outlets or cavities.

But farther than this, the child which is descended from parents who have themselves been subjected to like causes of physical debility,* must have an organization weakly from its very origin; must have inherited, to some extent, their debilitated constitution, and in consequence be much less able to resist the influence of the unfavourable conditions, to which it is exposed subsequent to birth.

This condition, it must be constantly borne in mind, has nothing to do with labour—as yet the child has undergone none: it has been allowed to run wild, half-clad and half-starved. It is of the utmost consequence to a correct understanding of the question of infant employment, that this preliminary state should be properly considered and correctly appreciated. It would, indeed, be a monstrous perversion of everything like justice, both to the masters and to the system of factory-labour, to burden the one with the moral responsibility of this prior condition, or the other with all the results for which it paves the way—a state, be it remembered, little likely to be ameliorated or improved, if the child were permitted to remain in idleness at home, since it would be still exposed to the same causes of physical and moral depravity. If, indeed, it could be shewn that abstinence from the mill would, to some very considerable extent, give it an opportunity for

* Vide Chapter VI., PHYSICAL CONDITION.

regaining during youth what it had lost during infancy, a powerful argument would be afforded for government interfering with what the abolitionists of this species of labour do not hesitate to term infanticide, and compare it to the Hindoo Gangetic superstitions, and the Chinese systematic child-murder. A perversion of words is, however, no argument; it may be ingenious, and serve party purposes, but it leaves the question at issue precisely in the same state in which it found it.

There can be no question that a child of nine or ten years of age, if it has been placed under circumstances to permit the healthy development of its organs, is capable of performing light work. If a child gets over its fifth year, even when it has shewn symptoms of slight scrofulous or rachitic affections, it generally pushes forwards, and does well, providing that it is properly nurtured, and is well fed and well clothed.

A child from an agricultural district, eight or nine years old, blooming with rosy health, full of vitality, is transported to a crowded town: its parents are compelled to leave it whilst they prosecute their work at the manufactories; it is exposed to miasmatic evaporations, shut up in a narrow street, its home is damp and cold, its food poor and badly cooked: in a few months, from the force of these circumstances, the Hebe-looking child has become pallid; its muscular system loses its tone; but it seldom goes down below this, its bony fabric having attained already so much firmness and completeness of ossification as to prevent it from yielding to any very great degree. In consequence, however, of the lessened efficiency of its digestive apparatus, it is much more liable to chronic diseases, glandular enlargement, irregular developments, at the expense of the whole system; and this train of circumstances is independent of labour, but is brought about simply by its change of locality, the altered habits of its parents, and its damp and unwholesome home.

The difference between the child born and subjected to these mischievous agencies, and the one born under more favourable auspices, in the influence of which it has lived till seven years of age, and then brought under the same train of mischiefs, as to diet, &c., when they are examined at nine years of age, is less striking than would have been previously supposed, from a comparison of data. Externally they present but few points of dissimilarity, whether in their physical structure, or in their morals: but a more attentive inspection will very readily enable the inquirer to ascertain that, in their several capabilities of bearing labour, or continued exertion, they are widely disproportioned, the preponderance of ability leaning vastly to the child whose prior healthy development had enabled its organs to make a decided progress.

In this instance its nutrition, for several years, had been unimpaired; its long bones had become in a great measure ossified; the whole increments of its system had been eliminated from a healthy source, through a series of vessels equally healthy and capable of performing their functions, and a degree of growth established, which no subsequent derangement could materially injure or deface. Not so the child whose destiny has fixed it here from birth: not only is its muscular system flaccid—supplied with thin watery blood; its mucous surfaces unhealthy; its glandular system deranged, but the groundwork upon which these are built, the frame upon which they are arranged, is incomplete in comparison. The bony fabric has had no opportunity for becoming properly solidified—contains as yet too large a proportion of cartilaginous matter; so that the child whose early years have been passed in healthy localities, and had home-tendance, is far more advantageously framed for work, and much less liable to the yielding of its osseous supports, than the child differently circumstanced.

Factory labour is a species of work, in some respects singularly unfitted for children. Cooped up in a heated atmosphere, debarred the necessary exercise, remaining in one position for a series of hours, one set or system of muscles alone called into activity, it cannot be wondered at—that its effects are injurious to the physical growth of a child. Where the bony system is still imperfect, the vertical position it is compelled to retain, influences its direction; the spinal column bends beneath the weight of the head, bulges out laterally, or is dragged forward by the weight of the parts composing the chest, the pelvis yields beneath the opposing pressure downwards, and the resistance given by the thigh-bones; its capacity is lessened, sometimes more and sometimes less; the legs curve, and the whole body loses height, in consequence of this general yielding and bending of its parts.*

These effects are necessarily produced only in instances where ossification has been retarded; and it has been shewn how the conditions of the child born and bred up in large towns tend to this state. With regard to the general health of the child, there is nothing in the labour to which it is subjected, which is likely to injure it; and it may be questioned whether cæteris paribus, it has not full as good a chance of its health being unimpaired in the mill, as if it were left at home, neglected as it inevitably must be. The inhalation of dust, &c., the effluvia from oil-steam, with

* "Had five hundred children, from five to eight years of age; work thirteen hours, one hour and a half for meals. I very soon discovered that these children, though well fed, and well clothed, and well lodged, and very great care taken of them when out of the mills, their growth and their mind were very materially injured, by being employed at these ages within the cotton-mills for eleven and a half hours per day. It is true that these children, in consequence of being well fed, well clothed, and well lodged, looked fresh to a superficial observer —healthy in their countenances; yet their limbs were very generally deformed, and their growth was stunted."—Evidence before Committee.

which the atmosphere of the rooms is infected, are in no other way injurious, than by shewing that ventilation is imperfect: of themselves they have no ill effects upon health, provided the consumption of the oxygen of the respired air is replaced by sufficient portions of fresh air. Odours, however disgusting—nuisances though they be—and though upon those unaccustomed to their fœtor,* they produce squeamishness, and a temporary indisposition for taking food—experience has proved are by no means deleterious. On the contrary, there are certain gases, which would appear to be even conducive to human health, though sufficiently disgusting. Soap-boilers, chandlers, stable-men, dealers in oil, curriers, tanners, are generally healthy men; and to some extent have appeared to enjoy an immunity from contagious diseases. Neither is the dust in the carding-rooms of cotton-mills so great as to produce any marked effect upon health. The temperature in which the infant portion of factory labourers work, is not so

* To prove that odours are not of themselves injurious to health, it may be remarked, that the oyster fishing affords as striking an example as can perhaps be found bearing upon the subject:—"It is a curious fact that these numerous couttôs (or heaps), each containing an enormous mass of oysters, all putrefy together on a narrow extent of soil, and emit the most detestable odours; yet the health of the precarious and crowded population gathered there is in no ways affected. During two consecutive years," says M. de Noè, "that I did duty at the fishery, I never saw a soldier of my regiment sick: Europeans and Sepoys all equally enjoyed good health."—*Memoires relatifs à l'Expedition Angluise de l'Inde en Egypte.*

high as to be actively pernicious, except in the first instance; nor does it extend beyond a certain point. When a child is first introduced into the mill, its temperature often produces a slight fever: this, however, does not go so far as to prevent the child working; neither is it at all contagious, but is simply the effect of temperature, and is preceded by considerable perspiration whilst it is at work. This of course produces some degree of debility and a paleness of skin, rarely recovered from; but the health does not continue to suffer from this cause; use soon reconciles the constitution to it. The mischief is, that in this climate, during certain seasons, it is difficult to keep up an increased temperature, without prejudicially interfering with ventilation; for when great numbers of individuals are congregated together, that portion of the atmosphere which is essential to life is rapidly absorbed, and a gas evolved fatal to animal existence. The proportion in which a room may be impregnated with this, although it may have no very visible immediate effects, yet it lowers the powers of life, and depresses the action of the heart, and so far aids the physical declension under which the child already suffers.* Great numbers of mills have however been examined, in which these disadvantages are in a great measure done away with, by a system of ventilation and cleanliness, as far as it is possible to attain, compatible with the nature of the operations carried

---

* Vide Chapter IX.

on in them; and where the amount of mischief resulting from the inhalation of floating particles, or of deoxygenated air, is so small, as to be worthy of little notice.*

Nothing appearing then in the condition of the labour to which children are subjected, of an active or positively injurious character, as far as physical health is concerned, the evils, if any, have their origin in causes foreign to it,—and these are, first, its continuance, and secondly, the prevention of all recreation and exercise necessary for sustaining or repairing their already shattered constitutions: and to these must be added, the moral effects of example, of independent earning and appropriation of wages, the evils of which have already been detailed.

The continuance of the labour for so many hours, is the one great physical agent for keeping up and increasing the inconveniences resulting from depraved nutrition. The whole burden, however, must not therefore be thrown upon factory labour; one half the mischief is done long before the child is engaged in it, and even where the period which it was engaged, deferred; the home and habits, of which it is the victim, are not those materially to assist in its regeneration.

* Nothing can more strongly mark the puerility which so often disgraces the opinions of scientific men, than the stress which has been laid upon the importance of a certain number of cubic feet of space being devoted to each individual in mills. The space, (generally about 1200 feet,) is more than enough, if ventilation is well performed, and is ridiculously small if this is imperfect.

To avoid misconstruction, it is highly necessary that the state of the child, preparatory to its being brought into the mill, should be correctly understood. To call out a number of children, and examine into their moral and bodily conditions, and to charge the whole of whatever is found bad in both or in either, upon the mill, directly (however truly, it may be charged with it remotely), would be manifestly unjust. "Suum cuique tribuito," should be strictly observed; the evils are enough without adding any thing beyond what fairly and properly belong to it.

It is beyond all question then, for it is abundantly *proved by physiological and pathological considerations*, that factory labour, continued for twelve or fourteen hours, is liable to produce certain distortions of the bony system, in consequence of the previous want of healthy growth; that it prevents proper and natural exercise; and that, in conjunction with a continuance of imperfect nurture, and want of domestic comforts, it keeps up an unhealthy condition of the digestive organs—leading or making the body peculiarly prone to take on a variety of chronic diseases, such as scrofula in all its protean forms, diseased joints, enlarged glands, &c.; and that it checks growth, partly by impairing the necessary supplies, and partly by positively lowering the height of the body.*

* The opinion of Dr. Baillie, one of the most enlightened and judicious physicians who have adorned the annals of British medicine, is coincident with the actual conditions as described

An inspection and admeasurement of 2000 children, taken indiscriminately from several large establishments, strikingly verified the accuracy of the foregoing results, deduced from reasoning upon the known functions and operations of the body in a state of health and disease. The children were stunted, pale, flesh soft and flabby; many with limbs bent, in most the arch of the foot flattened; several pigeon chested, and with curvatures in the spinal column; one hundred and forty had tender eyes, in a great majority the bowels were said to be irregular, diarrhœa often existing, and ninety shewed decided marks of having survived severe rachitic affections.

The female, from certain specific causes, is the one upon whom these bendings and curvatures are likely to produce the most serious consequences. The arch of the pelvis becomes contracted or made irregular in its outline;—the spinal column; when curvatures take place low down, encroaches

---

above, and is the more valuable, as having been entirely drawn from the general operations of the animal economy.

"I cannot," he says, "say much from experience, but I can say what appears to me likely to arise out of so much labour, from general principles of the animal economy. I should say, in the first place, that the growth of those children would be stunted—that they would not arrive so rapidly at their full growth—that they would not have the same degree of general strength—that it is probable their digestion would not be so vigorous as in children who are more in the open air, and less confined to labour; and that they would probably be more liable to glandular swellings than children who are bred differently."—Evidence before Committee.

on its dimensions, or forms such projections as effectually to block it up. To the natural difficulties and pains of parturition are thus superadded obstacles, not unfrequently fatal to the child, and in all cases exceedingly perilous to the mother, leading to the annual sacrifice of many lives, and an amount of human misery not conceivable by any one, but those whose avocations have led them to witness degrees of torture, which humanity shudders to contemplate.

The employment of children in manufactories ought not to be looked upon as an evil, till the present moral and domestic habits of the population are completely re-organised. So long as home education is not found for them, and they are left to live as savages, they are to some extent better situated when engaged in light labour, and the labour generally is light which falls to their share.* The duration of mill labour, from the natural state of the body during growth, and from its previous want of healthy development, is too

* "The labour in these mills is not perpetual labour, it is attention. The employment is not at all laborious—it is more a matter of attention than of labour."—*Evidence before Committee.*

"In those manufactories in which the time of work is not very long—where the children are not taken in at a very early age, and where the proprietors are men of enlarged minds, and possess some degree of benevolence, to induce them to look after the welfare of those from whom they derive their support and their wealth, the children are perhaps better off than in many other situations."—*Evidence before Committee.*

long, and masters would do well not to wait for legislation on the subject, but to dismiss their junior hands after eight or ten hours' labour, or it would be still better that they should not commence till eight o'clock, and should terminate at six in the evening.

There can be no question but that very considerable practical difficulties lie in the way of any extensive change as to the hours of labour,—difficulties, too, equally on the part of the masters and the men. It is doubtful if any legislative interference can be effective; but on the other hand, whether it may not most materially injure the future prospects of the labourers, and accelerate a fate already too rapidly approaching them. Still some modifications might be made to satisfy the claims of nature and humanity, contradistinguishing these from fanaticism and bigoted ignorance.

It can admit of little dispute, that by diminishing the length of time which the factory child now spends in the mill, some advantages would be gained by it physically; or, that the same advantages might be to some extent derived from periods of relaxation during the day. As it is, the perpetual necessity for attention, prevents any thing like bodily repose, so that, although no labour, in the common meaning of the word, is undergone, great exhaustion necessarily results. If any regulations, tending to do away with some of the evils attendant upon it, were introduced, and generally acted upon, it may be safely asserted, that the man who would abolish child-labour in large towns, does not understand the position in which children are there placed: or, if he does, he suffers himself to be led away by false notions of philanthrophy, and is no friend to the best interests of his species. The interior economy of mills has been so much improved, as to remove most of the obnoxious agents, which fell with such dreadful severity upon the parish apprentices, who first became their victims; and there is nothing whatever in a well regulated mill, directly injurious to life,—save only, the length of time spent there and its consequences. The real evil lies in the habits of the people themselves—habits, it is very true, generated by the system of factory labour, but decidedly not of necessity dependant upon it; and one half the mischiefs suffered by the children, are inflicted upon them prior to their commencing work.

The influence of infant labour upon the mental faculties, is a subject of less importance than is generally imagined. The children employed in mills, exhibit the peculiar intellectual traits which distinguish the town child from that of a rural district. They are acute and of quick intelligence; a circumstance arising from the greater variety of objects brought under their notice early in life, and to their being left to their own resources. it is not, indeed, improbable, that the exclusive attention which is required to one particular employment, may, in some degree, limit and confine the general capacity for acquiring diversified knowledge. This is, however, by no means proved to be the case; neither if it were, would it signify much, as this is a species of knowledge, which would not add to their individual happiness.

CHAPTER VIII.

HEALTH—RATE OF MORTALITY, &c. &c.

Political Economists—Poor Laws, &c.—Opinions as to Health, &c.—Rate of Mortality—Change of Employment, its Extent—Manufacturers, their Increase compared with that of Agriculturists—Improved Value of Life dependant upon a low State of Health—Change, &c. in the Habits of the People—Definition of Health—Mortality and Health among Savages—Mortality and Health in Agricultural Districts—Health in Manufacturing Towns—Low Condition of Vital Energy—Causes of this—Diseases, their Peculiarities—Almost Universally Chronic, and Compatible with Long Life—Table—Number of Patients Annually Relieved in Manchester—Medical Men—Comparison with other Large Towns—Difficulty as to Registration in Manufacturing Districts—Rate of Mortality in Agricultural and Manufacturing Counties—General Opinion—Dissenters, Proportion of, in Towns and Rural Districts—Causes of Apparent Disparity in the Rates of Mortality—Placed much too high in Towns—This Explained—Immigrations, their Effects—Infantile Mortality, its Extent—Crowding of Men into Towns—Influence upon Health—Mean Rate of Mortality in Manchester—General Summary.

WHATEVER opinion may be held as to the conditions of society, whether its physical condition is dependant upon its morality, or its morality dependant upon its physical condition, observation teaches that debasement in the character of the one, invariably leads to a corresponding declension in that of the other; that if a labouring community is forced into a state of pauperism, or one bordering upon it, morals, unless under circumstances as yet never seen, relax; and if, on the other hand, morals are from various causes neglected, the comfort and happiness of the mass are deteriorated in equal ratio.

The moral condition of the labouring population, in the manufacturing districts, is in itself deplorable; neither is the physical condition less so. Whether mutual reaction adds to their separate miseries, or whether each has its distinct and separate causes of misery, a little reflection will easily determine. If it were possible to amend the social and domestic habits—to produce cleanliness, attention to diet, proper clothing, and the decencies of common life, it is quite obvious that a vast improvement would at once take place in its physical comforts, and in its health. If it were also possible to amend the moral habits, to induce men and women to perform their relative duties, as wives, parents, and children—to make them religious, sober, provident, and economical, it is equally obvious also, that the same results would be produced.

Political economists have exercised their ingenuity in devising schemes and fables, and then have looked abroad, and endeavoured to explain the existing state of things, in accordance with their preconceived fancies. No wonder can be felt that tissues of absurdities should have usurped the place of common sense and observation. Clever and talented as some of these are, it is lamentable that they should pin their faith upon visionary conceptions, and confuse themselves and their readers with learned disquisitions upon population, national wealth, &c., &c., which have nothing whatever to do with the subject which is alone important to the well-being of a nation:—viz., the happiness, comfort and content, of each individual family, composing an integral portion of its strength and durability.

These mistaken views might be suffered to amuse those who are disposed to read them, did they aim at nothing higher; but when they are made the basis for arrangements intended to benefit and regenerate a moiety of the community, they present themselves in a more formidable point of view; and, by thrusting themselves upon the notice of men actually acquainted with the causes, conditions, and wants of this moiety, they deserve something beyond mere neglect.

The first step to be made is to improve the moral condition of the labouring population—without this nothing can avail it. It is in vain to expect that it will be made happier—that it will raise itself from the state of barbarous living into which it has sunk, without rousing into action its social and domestic affections. It is in vain to enable it to create wealth, which it cannot enjoy—in vain to tell it that it increases too rapidly, and to advise it to adopt "preventive checks." Its passions and instincts laugh to scorn such puerile nonsense, or turn the advice to still further poisoning the sources of domestic chastity. It is vain to tell it the scientific meaning words—to teach it to brew, or bake, or build, upon the most approved principles. This will not help it, for it is neither a philologist, a brewer, baker, or builder. It would be just as wise to write books for the Indian, whose habits are part of his nature, who knows not nor cares for creeds or formulæ, but who builds his faith in pleasure, animal indulgence, and selfishness. No: if benefit must be derived from this instruction, it must be preceded by teaching him his social duties, weaning him from his present course of life, and pointing out to him that he may be made happier by these means,—but the attempt to do this by mere intellectual tuition is alike futile and absurd.

The health and physical condition of the manufacturing population have their origin in, and are dependant in a great degree upon the perversion of their moral and social habits. They earn wages, which, with proper economy and forethought, would enable them to live comfortably, nay, in comparative luxury; as it is, they are in most of their domestic relations upon a level with the savage. Their reduction to this state of social debasement has been aided by certain mal-administrations of some laws, themselves of doubtful efficacy, in addition to their own singular depravity. The poor laws and bastardy laws have unquestionably acted, and are still acting, as premiums for immorality and idleness. This is produced, on the one hand, by rendering the poor man improvident, by furnishing a fund upon which he can retire; and, on the other, by destroying the chastity of the female, and thus ruining all those ennobling associations which should mark the intercourse between the sexes, and desecrating the most holy of all human engagements—that of marriage. That these evils are attributable full as much to a want of proper administration as to the inherent defects of these laws, cannot perhaps be denied; but laws which open a door so wide and tempting to profligacy and improvidence, must most assuredly be wanting in some of the proper attributes of legislative enactments. All reasoning *à priori* is done away with, and the evils of these measures is rendered abundantly apparent by the report of the "Poor Laws' Commission." To those individuals who had had opportunities of scrutinizing their operations, the evidence thus afforded was not necessary; but to men in high places, to political economists, who have hitherto been in utter ignorance of these things, a salutary lesson will surely be taught, and laws founded upon imaginary or illusory advantages will give place to arrangements based upon clear and decided knowledge—a knowledge acquired by practical and extended observation of things as they are.

The fallaciousness of the opinions deduced from the existing state of the population, held by men who have no real acquaintance with the subject, is very strikingly shewn by one circumstance. Mr. M'Culloch, and other writers of his school, from observing the improvement in the value of human life, and coupling this with its known rate of increase, have come to the conclusion that the health and comfort of the people at large must have materially improved. Joining this fact to a consideration of the change which has been going on in the occupation of the people, namely, their rapid conversion into manufacturers,—a conversion so rapid and extensive, that whilst the entire population has increased from 1801 to 1831 rather more than 50 per cent., in the manufacturing towns and districts this increase has advanced 140 per cent.,—they suppose that the increased longevity of the whole must indicate that manufactures are decidedly healthy,—notwithstanding since steam became the moving power, they were of necessity confined to particular localities for the convenience of fuel, and crowded into towns and populous districts for a supply of hands.*

According to Mr. M'Culloch, the average rate of mortality in 1780 was 1 in 40; 1810, 1 in 53; 1820, 1 in 57; and, it may be added, in 1833, 1 in 60.

Mr. Finlayson's calculations enabled Sir G. Blane to draw up the following table, which shews very distinctly the comparative mortality at two distant periods. These calculations, it must never

* Vide Mr. Senior's Lecture on Wages.

be forgotten, are to be looked upon rather as approximations to reality than reality itself: still they are highly valuable.

Ages.	Mean duration of life in 1693	in 1789	The increase of vitality is in the inverse ratio of 100 to
5	41.05	51.20	125
10	38.93	48.28	124
20	31.91	41.33	130
30	27.57	36.09	131
40	22.67	29.70	131
50	17.31	22.57	130
60	12.29	15.52	126
70	7.44	10.39	140

The changes which have taken place in the different classes of the community are equally curious, as shewing the progress of manufactures from 1811 to 1821. These changes have been reckoned upon ten thousand of each class.

	Agriculture.		Manufactures.		Unproductive as professions, &c.	
England . . .	Decrease	168	Increase	175	Decrease	7
Wales	do.	555	do.	63	Increase	422
Scotland . .	do.	211	do.	34	do.	178

Taking the counties separately, the increments of agriculturists are 1437, for trade and manufactures 10,658, a proportion seven times greater than the first.

According to Mr. Marshall's analysis of the population in 1821 and 1831, it appears that in the manufacturing districts it had increased one-fourth, and in the agricultural only one-eleventh.

The rapidity of increase in the manufacturing population, and its extraordinary growth, compared to that engaged in agriculture, is indicated by the fact that, in 1801, it was calculated as six to five, in 1821, as eight to five, and in 1830, as two to one. This transition, for so it must be called, is still farther elucidated by the following table, drawn up from the report of the Committee of the House of Commons, on manufacturers' employments, in July 1830, and from the census of 1831. In a period of thirty years, the following increase per cent. appears in the manufacturing towns.

Places.	1801 to 1811.	1811 to 1821.	1821 to 1831.	1801 to 1831 } Total.
Manchester ..	22	40	47	151
Glasgow.....	30	46	38	161
Liverpool....	26	31	44	138
Nottingham ..	19	18	25	75
Birmingham..	16	24	33	90
Great Britain .	14.2	15.7	15.5	52.5

The population of Lancashire, which is the great centre of the cotton trade, in 1700, was 166,200; in 1750, 297,400; in 1801, 672,731; in 1811, 828,309; in 1821, 1,052,859; in 1831, 1,335,800.

When the increase is compared with that of an agricultural country through the same periods, the disparity will be strikingly seen. Norfolk for example, in 1700, 210,000; in 1750, 215,000; in 1801, 273,371; in 1811, 291,999; in 1821, 344,368; in 1831, 390,000; an increase of one and three-fourths, whilst Lancashire in the same period has added to its population nine fold.

These details are amply sufficient to shew the rapid growth of the population engaged in manufactures when compared with the general increase of the inhabitants of the kingdom at large, and with that division more immediately connected with agriculture. The numerical disproportion which is steadily progressing shews very clearly, when joined to the diminished rate of mortality evidenced by the general increase, that there can be nothing in the processes of manufacture, necessarily destructive of human life. On the contrary, it may be asserted that the improvement in its average duration, is intimately dependant upon the conversion of the bulk of the population from agriculturists to manufacturers.

Paradoxical as it may sound, it by no means follows that because the duration of life is extended, the people are a more salubrious race than their forefathers, whose lives averaged hardly 1,35. It is indeed true, that many of the fatal diseases which formerly at certain times nearly depopulated whole provinces, are no longer in active and extensive operation: plague, sweating sickness, petechial fevers, small pox, the scourges which during their periodical visitations destroyed hundreds of thousands, are themselves gone to the "tomb of the Capulets." This salutary change has been brought about not by improvement in the art of medicine—though very great and very admirable discoveries have been made, and, above all, the treatment of disease has been much simplified and become more rational—but has been in a great degree produced by the alteration which has been gradually going on in the habits of the people, and their modes of living. The rush-covered floor, generally unflagged, a receptacle for the filth of weeks—the animal diet, the ale drinking, the popular sports and seasonal celebrations, each in their way induced a state of health unfavourable to longevity. The narrow and crowded streets, the small and low rooms, with their contracted windows and thick walls, nearly dark and badly ventilated, the want of proper drainage, and half occupation were all predisposing causes for the generation or propagation of mischievous contagions. War, famine, want of medical aid amongst the poor, were other abundant sources of destruction.

Notwithstanding these drawbacks, the population of Great Britain enjoyed much better health than it does at the present period, and this too, notwithstanding the diminished rate of mortality. Taken as individuals, they were more robust, fuller of organic activity, enjoyed in much higher degree, the feelings of existence; but, in consequence, their diseases were of a much more acute character, and infinitely more fatal in their tendency.

Health may be defined that condition of the body, in which all its functions or operations go on without exciting uneasiness. Nature no doubt meant, that as far as mere existence is concerned

—that is, that every thing connected with the taking of food, and its necessary changes, constituting nutrition, and every thing connected with sensation—the link which binds man to the material world around him—should be a source of pleasurable feelings, or, at all events, that it should be unattended by pain; were it otherways, mankind would be stretched upon the rack of its own sensibilities. Whatever produces an opposite state of things, must be looked upon as causative of disease, or that condition of body, in which some of its functions are so far deranged as to interfere with personal comfort, inducing uneasiness or pain, or proceeding farther, and deranging and disordering the action of parts, so as to interfere with the operations essential to life, and, in the end, producing its extinction.

The rate of mortality among savage tribes, and in states just emerging from barbarism, is universally high; one in thirty, as far as data can be collected, being the average. The health of individuals composing these tribes or states, is nevertheless, according to all travellers, exceedly vigorous, and their frame generally robust, indicated by the great fatigue they are capable of enduring, the severe wounds from which they rapidly recover, the rarity of chronic diseases amongst them, and the few decrepid and premature old men, who burden their huts or wigwams. Infanticide and voluntary abortion, neglect in the early periods of life, the little estimation in which female children are held, and the want of remedial means of even doubtful efficacy, destroy vast numbers shortly after birth. A great deal is often said as to the hardihood of the Indian mother, and the vigorous nature of her offspring, evidenced by her easy delivery generally unassisted and uncared for, and from the circumstance of her slinging her infant on her back, and almost immediately joining the march, or pursuing her household or field labour. But if a list of children who perish from want of proper attention could be obtained, it would present a frightful picture of infantile mortality.* The habits of women in civilized countries, are repugnant to the simple forms of savage life; and in most cases, the infant is tenderly nurtured, and, in consequence, vast numbers are reared, which, under a different regime, must have inevitably perished; and this is one very important item in the account of the comparative mortality in the two conditions. Again, the adult male population in savage nations, is seldom at peace—war being the maximum of their animal existence and happiness. This is especially destructive, by cutting off in their prime the very flower of the men, and laying the remainder of the tribe open to exterminating aggression. It is, indeed, rare to find an adult labouring under

* Vide James's Expedition to the Rocky Mountains; Humboldt's Travels; Cruze's New Zealand; Dobrizhoffer; Collin's New South Wales; Franklin's First and Second Journies; Parry's Voyages; Vaillant, Bruce, &c. &c.

sickness; nine-tenths being carried off either by war or accidents in the chase, or falling victims to acute inflammatory disease. Famine is another powerful agent in the destruction of tribes, whose existence depends chiefly upon the produce of hunting or fishing, or upon the imperfect cultivation of a very limited portion of ground; and this again exposes them to contagious or epidemic diseases, which often sweep away a whole people. The rate of health is, however, high, and their bodily vigour well known to all inquirers, frequently, indeed, presenting splendid examples of almost animal perfection.

The condition of man in a state of civilization, differs of course very widely, in some of these particulars. He is better lodged, better clothed, and exposed to fewer and less important variations. The agricultural districts, where labour is moderate and food abundant and nutritious, the labourer has, in many points, a bodily constitution closely resembling that of the savage. He possesses great physical development, high health, uninfluenced by intellectual irritability, and is, consequently, open to the attack of acute diseases, which soon destroy him, when brought under their influence. Fevers, inflammations in important viscera, resulting from deranged circulation in a system of blood-vessels, gorged with nutritive fluids, are rapidly fatal, without the most prompt and energetic measures for their relief, and which he is usually unfavourably placed for obtaining.

An examination of the lists of patients of a country practitioner, in a district exclusively agricultural, will verify the correctness of these observations; the majority of deaths resulting from inflammation of the chest, head, and abdominal or pelvic viscera, or the consequence of accidental violence, and exhibiting very few chronic diseases, except the anomalous ones amongst females.

Thus, the state of health is high in agricultural districts, and amongst savage and semi-barbarous tribes. It is, indeed, very common to hear individuals so circumstanced, declare, that they have never known a day's sickness, or been plagued with ache or pain, when a rapid disease seizes them, and their vigorous health and great vascular activity become the agents of their destruction,

In comparing this state of things with the correspondent ones in manufacturing towns and districts, a very striking contrast is observed. In the former case it may be truly said, that life is physical enjoyment, and disease hasty death; in the latter, that life is one long disease, and death the result of physical exhaustion.

The population crowded into the large manufacturing towns, which have sprung into being almost with the rapidity of Aladdin's palace, and where their growth has preceded all efficient police regulations, is exposed to many causes tending very powerfully to depress its vital activity. Unpaved and unsewered streets, ricketty houses, huddled into heaps, undrained, unprovided with needful conveniences, badly ventilated, and crowded with inmates; the habits of the population itself, its improvidence, its neglect of domestic comforts, its indulgence in dram-drinking, its general immorality, its thin and innutritious diet; and these, joined to the peculiarity of their labour, continued unremittingly for twelve or fourteen hours, cooped up in a heated atmosphere, and debarred from the cheering influences of the green face of nature, and of fresh air, and, finally, deprived of all recreation* in open and salubrious situations—all these are agents for lowering the tone of the system, and produce a series of diseases, widely dissimilar from that which has been explained as proving so destructive to a population placed under opposite conditions.

In these towns, disease generally assumes a chronic type; its progress is slow, and often interferes but little with the proper functional actions which are essential to life. Neither, in many instances, does it, of necessity, shorten its duration; but rather, by keeping the standard of vital energy somewhat below par, it abstracts the system from the impression of more fatal affections, which kill by disturbing the circulation. To illustrate this, the following table, including 5,833 patients, who came under the care of one medical gentleman, attached to the Manchester

* Vide Chapters III., IV., V., VI., SOCIAL AND PHYSICAL CONDITION.

Infirmary, has been drawn, specifying the nature of the diseases in as simple a manner as possible. These cases occurred in the four years from 1826 to 1830.

Disease	Cases	Disease	Cases
Inflammation of the Brain, &c.	6	Secondary Syphilis	48
Inflammation of the Tonsils, &c.	41	Bronchocele	5
Inflammation of the Bronchiæ, &c.	31	Dentition	5
Pleurisy, &c. &c.	80	Abortus	1
Affections of the Liver, &c.	32	Scrofula, Common	18
Inflammation of the Bowels, &c.	26	Hæmorrhages	62
Inflammation of Bladder, &c.	17	Dyspepsia, &c.	203
Rheumatism	569	Jaundice	19
Intermittent Fevers	47	Cholera	45
Remittent Fevers	11	Diarrhœa	191
Common Continued Fever	861	Dysentery	323
Common Catarrhal Fever (Colds)	550	Constipation	755
Measles	8	Colic	72
Scarlet Fever	24	Verminatio	44
Small Pox	19	Hemorrhoids	15
Chicken Pox	2	Hypochondriasis	28
Erysipelas	26	Headaches	51
Purpura	4	Apoplexy	1
Nettle Rash, &c.	47	Paralysis	50
Simple Cough	640	Epilepsy	25
Hooping Cough	21	Hysteria	20
Asthma and Difficult Breathing	297	Chorea	27
Loss of Voice, &c.	8	Convulsion (Children)	3
Consumption	228	Palpitation	12
		Anginæ Pectoris	1
		Nervous Pains	26
		Diabetes	4
		Ischuria	8
		Dropsies	110
		Amenorrhœa, &c.	47
		Leucorrhœa, &c.	14
		Poison	1

This singular table very clearly and satisfactorily shews the prevailing character of diseases amongst the operative population, medically considered. The great majority of these are compatible with a very extended life; few are fatal of themselves, and still fewer with the aid which is liberally afforded them. The acute diseases are exceedingly small in number, when compared to the whole; and this is the more worthy of notice, since they are precisely those which are the most likely to come under the observation of the medical officers attached to this admirable institution. It will be remarked that the cases of dyspepsia, constipation, and other affections dependant on derangement of the digestive apparatus, are more than one-third of the entire cases. This most distinctly indicates the vast amount of these peculiar derangements; for great as their number is in this one table, they must have been extreme cases to bring them under notice. Another peculiarity distinguishing the list is, that one-fifth of the number is made up of coughs, &c., while scrofula and consumption, in its varied forms, are exceedingly limited.

The number of patients annually receiving advice and medical assistance, at the various munificent charitable institutions in the metropolis of the manufacturing world,* will afford another valuable document to estimate the rate of health enjoyed by the community; all, however, that are relieved here, are not of the class of operative manufacturers. The number of individuals resorting to these charities cannot be less than 30,000

* Manchester.

per annum,* an immense proportion out of a population of 240,000. There are, in addition to these, 133 surgeons, 26 physicians, 76 druggists and apothecaries, making a total of 235, all and each of whom procure a livelihood by ministering to the ailments of the different classes of society. Besides these, there are a host of quacks, at all times a flourishing order, sellers of patent medicine to a great amount annually; those too who adhere to domestic medicine, and lastly, and by no means a small item, the numbers who resort to many physicians and younger surgeons for gratuitous advice, perhaps not less than 2000 in the year. To this must be also added, the female patients of the Lying-in Hospital, amounting, at the present period, to 5000 more, who are supplied with medicine for diseases incident to gestation, accouchement, and subsequent recovery; also their infants till arrived at a certain age. Taking all these together, it may be inferred, very safely, that 3-4ths of the entire population require medical aid annually; a vast proportion certainly, and in all probability greater than what exists in other large towns more favourably placed in some respects than Manchester.†

---

* In 1831, patients admitted at the Royal Infirmary, 21,196; House of Recovery, 472; Ardwich and Ancoat's Dispensary, 3,163; Workhouse, 2,100; Children's Dispensary, 1,500; Lock and Eye Institutions, 1,500? Chorlton and Hulme Dispensaries, 1,000?

† If a judgment may be formed of the number of sick from the number of medical men, Manchester is very unfavourably placed in this respect; though the criterion is far from being a

It might be supposed, in looking at this formidable list of disease, with the number of charities, and the array of medical men devoted to its alleviation, that death would sweep away a great proportion of a people possessed of so little physical energy; that, consequently, manufactures, with their necessary crowding together of men in towns and mills, would be singularly injurious to the average duration of human life. To some extent, doubtless, this is so, but a little examination will shew the fallacy of judging of this from the state of health enjoyed or suffered by communities.

The very imperfect registers of births and deaths—so generally complained of in Great Britain—exist in full force in the manufacturing districts. It is indeed utterly impossible to form any estimate even approaching to accuracy as to the rate of mortality in any given amount of population. The increase depends, very materially, upon immigration from surrounding neighbour-

---

fair one. Taking the population of London at 1,200,000, in 1821—

There were	Physicians	174	or 1 to 7,000
	Surgeons	1,000	or 1 to 1,200
	General Practitioners	2,000	or 1 to 600
	Druggists, &c.	300	or 1 to 4,000
Paris at 800,000, in 1821—			
	Surgeons	128	or 1 to 6,000
	Physicians	600	or 1 to 1,300
	General Practitioners	180	or 1 to 4,450

London Total 3,474 or 1 Medical Man to every 345 inhab.
Paris ....1,310 or 1..............................900 ——
Manchester... 235 or 1..................................121-3 inhabitants, in 1831.

hoods, to which again many of those who die are removed for interment. Great numbers of interments take place in the towns from the out-townships, so that the registries which do exist, exhibit burials only, without any particular specification which can be depended upon. Then as to registries of baptisms, as data for births in a particular town or locality, these are no guide whatever, looked at generally. A very great proportion of the baptisms at the established churches, where alone any tolerable register is kept, being from out districts. The chief difficulty, however, lies in the circumstance of the majority of the inhabitants engaged in manufactures being dissenters; and from the fact, which is much to be regretted, that amongst them, as a body, it is out of the question to collect satisfactory statistical details. Again, immense numbers of the children born to parents in the lowest grades of society, are never baptized at all, and if these perish, they are buried in the free burial grounds without the slightest attention to registration. The population is an exceedingly fluctuating one, differing widely from that in agricultural districts, and presenting insuperable obstacles to the collection of accurate details. When all these things are taken into consideration, it must be quite obvious that with so many sources of error in the way of partial examination, it will be safer to adhere to the general rate of mortality for the whole kingdom, which can be ascertained with considerable accuracy, and establish deductions taken from that.

It has been universally asserted that the rate of mortality, in counties largely engaged in manufactures, was much higher than in agricultural counties. The following table shews the generally received opinion upon this subject.

Manufacturing Counties.	Deaths to population.	Agricultural counties.	Deaths to population.	Wales.	Deaths to population.
Middlesex	47	Sussex......	72	Anglesey...	83
Chester	55	Monmouth.	70	Brecon......	67
Lancashire....	55	Gloucester.	64	Cardigan ...	70
Yorkshire.....	60	Suffolk......	67	Pembroke..	83
Stafford	56	Wilts	66	Carnarvon..	69
Warwickshire	52	Hereford ...	63	Glamorgan.	69
Average	53		67		73

This table, which has been selected in consequence of having been drawn up for a very opposite purpose, and which gives the extremes, might at once seem decisive as to the question of mortality. When fairly explained, its disproportionate results will be materially changed.

In the agricultural counties, as has been before remarked, the population is a fixed one, and, generally speaking, one attached to the church, and exceedingly scrupulous in every thing relating to deaths and baptisms. The following table will verify this :—

Manufacturing Counties.	Established Churches.	Dissenting congregations	Agricultural Counties.	Established Churches.	Dissenting Congregations.
Middlesex.....	233	289	Sussex......	300	87
Chester........	145	153	Monmouth.	118	72
Lancashire....	287	504	Gloucester..	290	117
Yorkshire.....	809	1,019	Suffolk......	486	132
Stafford........	178	213	Wilts........	274	129
Warwick......	209	108	Hereford....	201	49
	1,861	2,286		1,669	586

Nor is the numerical disproportion all that must be borne in mind in carrying on the comparison. In the manufacturing counties and towns, the dissenters' congregations are vastly more numerous than those of the established church, whilst in the agricultural counties they are generally very small, and placed in remote situations.

This single fact of the difference existing in the religious forms between the agriculturist and the manufacturer, will show how much more accurate are the means of ascertaining the particular condition of the one over those of the other, and will explain one cause why all the tables of mortality exhibit so great a preponderance in favour of the agriculturist, greatly indeed above what in truth belongs to him. It is not denied that the rate of mortality is higher in towns and manufactories, but the disparity it is asserted is much less than is generally conceived. The immigrations which are continually going on from the agricultural districts to the manufacturing ones, consist of those families who would be precisely the parties to swell the bill of mortality in their own parishes by their poverty, and are little likely to improve that of the situation in which they settle themselves. These families are generally those that have numerous children, whose parents have no means of supporting them by agricultural labour. These therefore appear in the lists of births, but no further. Again, the extensive immigrations into these districts are uniformly driven there by want, and often suffer very severe privations before they can obtain work, and many of them perish in consequence, especially the children, from the change of locality, and hence swell enormously the deaths in the towns; but these have nothing to do with the fixed population. Farther still, great numbers of these families bring with them the older members, who have hitherto lived in open and healthy situations, favourable for the prolongation of life, but who sink at once be neath the depressing influence of their new abodes. When these several drawbacks are taken into account, the mean mortality naturally incident to manufacturing districts will be considerably lowered, for it would be a manifest injustice to load these with the imputation of a destruction which is perfectly foreign to them. But still the mortality will not be brought down to that in agricultural districts, in which it has been shewn the elements for fatal disease are much more rife, and where consequently a rate of death should be exhibited according to this mode of viewing the question as high, if not higher than in the opposite case. This remains to be explained.

In that portion of this work which has been devoted to the examination of the social and domestic conditions of the manufacturing population, it has been demonstrated how closely in some respects these approximate to the modes of life in savage * and half civilized states or tribes;

* Vide Chapters V. and VI.

and in a previous portion of the present chapter it has been said, that amongst people so situated, one great cause of the excessive contraction in the average duration of life, was to be found in the number of infants who perished shortly after birth, and in the prevalence of infanticide and voluntary abortion. Neither is the manufacturing population huddled into towns backwards in these respects;--thus adding another feature to their similarity to barbarism in its most odious forms. It has already been stated that the majority of children delivered in Manchester, the offspring of the poor, are brought into the world by the aid of public charity—that these children are abandoned by their mothers in a few days after birth to the care of strangers; that they are of necessity badly attended to, badly fed, exposed to cold, or are crammed with patent medicines or dosed with gin; and the consequences are that more than one-half of all children born to the lower class perish before they have completed their fifth year This appears upon the face of the bills of mortality, and their imperfection has been pointed out;—and in this case they only serve as an index to the enormous extent of infantile mortality—an extent equal if not greater than what exists in the most uncivilized nation on the face of the globe. This too, when compared with the average number of deaths before ten, explains at once one reason why the general rate of mortality is so high in these situations.

Manufactures have also the inevitable effect of bringing men together in masses, and hence the districts where they chiefly prevail may be looked upon as one vast town; and this is another cause of the increase of mortality, and one too which, *de facto,* has nothing to do with the town being a manufacturing one. The rate of mortality, as far as can be ascertained, is about 1 to 35 to Manchester; and though this at first sight appears excessive, yet, when compared to other great towns, it ceases to be remarkable.* In Paris it is as 1 to 32; in London, 1 to 34; in Birmingham, 1 to 40. In making these calculations, and in reasoning upon them, it must never be forgotten that, from the rapid growth of the manufacturing

---

* From a parliamentary paper, containing a return of the number of burials occurring annually in Manchester, from 1821 to 1830, and returns obtained by the Board of Health, the following table has been constructed:—

Year.	Interments of churchmen.	Interments of dissenters.	Total of interments.	Population	Rate of mortality.	
1821	1,561	1,726	3,287	152,683	46.45	
1822	1,285	1,044	2,329	156,663	67.223	
1823	1,585	3,230	4,815	160,664	33.36	mean
1824	1,428	3,219	4,647	166,117	35.74	rate:
1825	1,398	3,539	4,928	173,083	35.12	35.22
1826	1,548	3,804	5,352	180,052	33.64	
1827	1,604	3,235	4,839	186,462	38.53	
1828	1,615	4,106	5,721	192,874	33.73	
1829	1,479	3,719	5,198	201,691	38.80	
1830	1,590	4,383	5,973	212,913	35.64	
1831			6,736	224,143	33.27	

This table shews, too, very distinctly, the preponderance of Dissenters numerically, which is even considerably greater than the proportion it exhibits.

towns, and from the fluctuating and uncertain nature of their population, they were built without the slightest regard to health or convenience; and the account given in Chapter IV. will shew in how lamentable a condition Manchester was found in 1832, in every thing having reference to the health, decency, or comfort of its inhabitants.

When these various concurrent causes, adding to the mean mortality of manufacturing counties and towns, are taken into due consideration, and proper weight attached to them, an explanation is afforded why, in all tables of increase and decrease of population, these counties always seem very unfavourably placed; but these concurrent causes have most clearly nothing whatever to do with manufactures as an occupation, and their influence upon health and mortality.

The details in the beginning of this chapter have shown, that during the last fifty years the entire population has been undergoing a rapid conversion from an agricultural to a manufacturing character: that in 1800, it was calculated that manufacturers were to agriculturists as 6 to 5; in 1825, 8 to 5; and in 1830, as 2 to 1; that if this examination is carried farther back, the tables are turned, and agriculturists have the numerical majority. Beginning in 1780, they were about equal; 1760, 6 to 5; 1740, 8 to 5; 1700, 2 to 1; and so on, till the great body of the inhabitants were exclusively devoted to agriculture; that at the period when this was the condition of society, the average mortality was fully 1 in 35; in 1780, when manufactures had received their first great impulse, 1 in 40; in 1810, when the bulk of the population was engaged in them, 1 in 52; in 1820, 1 in 57; and in 1830, 1 in 60. Hence it appears that during the transition of employment, the mean duration of life has been steadily improving, and that at the present time, its value is double to what it was in 1700. How manifestly unjust—how manifestly absurd, to declare that manufactures are injurious to human life. On the contrary, it may be said that, was the population a fixed one, and one sober and moral in its character, it would shew a rate of mortality infinitely superior to that in agricultural counties.

Life, however, though not necessarily shortened by manufacturing occupation, is stripped of a most material portion of that which can alone render it delightful—the possesion of health, and those who are engaged in it may be said to live a protracted life in death.

CHAPTER IX.

FACTORY LABOUR IN GENERAL—PECULIAR DISEASES, &c.

Wages—Bodily Labour—Attention—Temperature—Effects of Heat—Vegetable and Animal Decomposition—Difference of these upon Health—Effect of Continued Labour in High Temperatures—Smell of Oil, &c.—Inhalation of Dust—Knife Grinders, &c.—Dust in Cotton Spinning—Effects of Breathing this Dust—Does it Produce Organic Disease?—Dust Proceeding from Filing of Steel—Polishing Marble, and its Difference from Cotton Flue—Habit—Bronchitis—Comparison of Mortality between Weavers and Spinners—Error of the Received Opinion as to the Injurious Effects of this Dust—De-oxygenation of the Atmospheric Air in Mills—Production of Carbonic Acid Gas—Qualities of the Gas, and the Consequences of Breathing it—Its Specific Gravity, &c.—Change of Temperature—Summary of their Conditions—Scrofula—Indigestion—Fever—Consumption, &c. &c.

Before proceeding to examine the diseases to which the manufacturing population are peculiarly exposed, it will be useful to take a survey of factory labour in general, and the conditions (if any) which render it so noxious to health, as is universally imagined.

The persons engaged in cotton mills earn much higher wages* than most other classes of labourers.

* Bricksetters, builders, joiners, painters, &c. &c., earn, it is true, higher wages per day, and their labour never exceeds ten

Spinners of fine yarn can earn 25s. to 30s. per week; coarse spinners, chiefly women, 18s. to 21s.; children of nine years of age, and upwards, get from 3s. to 4s. 6d.; weavers, chiefly young women and girls, 10s. to 16s.; and various other rates of wages, none below 3s., are paid to the batters, pickers, carders, stretchers, doublers, reelers, makers-up, warpers, winders, &c., which form the complement of workers in the mills, as at present constituted. The average rate of wages for each individual may be fixed at 10s. per week; this, of course, includes the children, who in all instances form a great proportion of the hands. This rate of payment is very nearly the same as that which was obtained in 1816. The usual hours of labour are from five o'clock in the morning to seven o'clock in the evening, half an hour being allowed for breakfast, one hour for dinner, and half an hour in the afternoon. The time of labour has also undergone very little change during the last twenty-six years.

With respect to the quantum of bodily labour gone through by the operatives;—generally speaking it is exceedingly small. The work of spinners and stretchers certainly is laborious at times, and from its length, twelve entire hours, it is necessa-

working hours. A day's wages for these men may, as an average, be stated at four shillings. The great drawback is, that the demand for their labour is fluctuating and uncertain, and at some periods of the year they are in complete idleness. The manufacturers employment is invariable.

rily accompanied by great exhaustion. In these processes the hands are hardly an instant at rest, except whilst their mules are doffing, in which they sometimes assist; and however hardy or robust a man may be, this incessant exertion is overpowering. But in all other cases, as in the carders, rovers, winders, piecers, weavers, &c., very little manual labour is required—none, indeed, worth speaking of. The only fatigue suffered is that brought on by the necessity for unintermitting attention—a fatigue more injurious to health than that induced by physical exertion, if not excessive.

The mean temperature in the factories may be taken at 70° Fahrenheit. In dressing and finishing, a much higher temperature is necessary, in the present mode of going through these operations. It is, however, likely that certain changes will be introduced, very shortly, which will very considerably modify this. Inquiries amongst the dressers have shewn that, as a body, they are by no means unhealthy, when their habits are moderately regular. They earn excellent wages, perspire during their work very copiously, and in place of supplying this waste by proper food, drink largely, a fact quite sufficient to account for their occasional illnesses, and general emaciated appearance. Temperature in itself is in no wise injurious to health, provided it is not joined with close and imperfectly ventilated situations. The difficulty in the mills has always been the keeping up the requisite degrees of heat, with a free admission of fresh air; and some years ago the system was decidedly noxious. By means of steam-pipes traversing the rooms in various directions, and giving out certain known quantities of heat, the mischief has in a great measure been done away with, and in the best regulated mills the temperature, though considerable, is in no farther degree oppressive or injurious than as its relaxing muscular tonicity. Hot climates may be, and are healthy, when cleared from vegetable undergrowth, and free from swampy soils. Humboldt and other travellers and observers have examined this point quite sufficiently. Heat generates miasmata fatal to human life, wherever moisture, joined, as it universally is in tropical countries, with rank vegetation, exists. The littoral districts of New Spain, the cedar swamps bordering the great South American rivers, the jungles of Africa, and of the torrid zone in Asia, bear evidence how deadly an influence may be produced by vegetable decomposition, &c. &c. Beyond these, however, perfect salubrity is quite compatible with a very elevated temperature. A wide distinction must be here made between animal and vegetable decomposition—a distinction little understood, and still less attended to. The gases which are evolved during the putrefaction of animal substances do not appear, from experience, to be fatal to life, but, on the contrary, in some respects, and to some extent, seem rather conducive to health. In a note which will be found in the chapter on Infant Labour, some remarks have been made on this subject; and in illustration, M. de Noe's account of the Oyster Fisheries has been quoted. Here, although millions of oysters are putrefying under a burning sun, in the very midst of a dense and promiscuous mass of human beings, filling the atmosphere with a most intolerable stench, sickness is hardly known. In the process of grinding bones in this country for manure, a smell the most dreadfully offensive attends upon the operation; yet the men, who are constantly inhaling this odour, are exceedingly healthy. Butchers, tripe-men, tanners, candle-makers, &c. are all exposed more or less to animal matter, in various stages of decomposition; ostlers, night-men, are under similar circumstances, and yet, all things considered, are far from being unhealthy. Not only this, but many nations devour ravenously putrid meat, and, in our own times, several species of animal food are esteemed unfit for the table till putrefaction is very considerably advanced; and yet these are considered as wholesome articles of diet.

It is often said that factory labour must be exceedingly prejudicial to life, in consequence of the crowding together numbers of individuals in one room, or one mill, from the effluvia which proceed from the bodies of these individuals—none of them, probably, very cleanly, and all in a state of perspiration, more or less profuse. This opinion is not borne out by facts, where the crowding is not extreme. Many prejudiced persons have drawn a parallel between the factory and the Black Hole of Calcutta, and have suffered their imaginations to run riot, in colouring a series of dismal pictures, similar to that so forcibly and graphically described by Mr. Howell. There is no similarity in the two cases whatever;—therefore deductions drawn from the one are not applicable to the other.

The consequences of labour carried on for twelve hours, without intermission, in a heated atmosphere, which slightly accelerates the action of the heart, are, that the muscular system becomes languid, and a degree of irritability is produced, which is equally exhausting to the functions of the nervous system, both of which excite feelings of great discomfort and depression. Still this is not disease. It is true that it is a condition closely bordering upon it, but it neither is disease, nor would it lead to disease, if the operative had a home stored with domestic comforts, which he could well afford, and had habits which led him to devote all his spare time to their enjoyment. What is the fact? His home itself is a very sty, abounding in all the elements for exciting and perpetuating disease. In place of seeking, by nutritious food and rest, to dissipate his languor and exhaustion, he flies for temporary relief to the excitement consequent to drinking ardent spirits—the very mode to add to his other evils, and to develope the seeds of disease lurking within him, generated partly by the nature of his occupation, and partly by his want of proper diet.

The exhalations from oil, &c. in a heated atmosphere, though sickly in their impressions upon strangers, there is no reason to believe are injurious to health; probably, indeed, the reverse, as it is a well-known fact that oil-men, and oil-porters, have enjoyed a singular exemption from the attacks of plague, and other epidemic and contagious diseases. So far has this been the case, that as a protection during the visitation of these scourges, it has been recommended to wear oiled dresses, or to anoint the body with oil several times a day. The copious use of oil for the purpose of lessening friction does away with the notion, that this going on so extensively in mills, between metallic bodies, might give rise to a state of atmosphere something, though remotely analogous, to that breathed by knife-grinders and pointers of needles, and which is exceedingly destructive. Every precaution, however, is taken to lessen the amount of this friction, which rapidly destroys the parts upon which it acts, and which, without these precautions, would speedily ruin the machinery.

Several writers who have devoted their attention to the influence of particular occupations upon health, have noticed the effects of inhaling particles of solid matter into the bronchi, &c. &c.

No question exists, that when an atmosphere is surcharged with foreign matter, so finely pulverized as to float easily, the breathing such an atmosphere must be more or less injurious. The extent of the mischief which will result from this, however, must depend very considerably upon the nature of the inspired particles. The vivid sketch given by Dr. Knight,* has shewn very decidedly that all the operations of steel grinding and polishing produces a species of asthma and phthisis in those engaged in them, which generally proves fatal early in life. The researches of Patissier,† in relation to cotton-spinners, led him to remark, "These workmen constantly inhale an atmosphere loaded with very fine cotton dust, which excites the bronchi, provokes cough, and maintains a perpetual irritation in the lungs. They are often obliged to change their employment in order to avoid phthisis." Similar observations have been made with regard to other branches of trade, such as stone-masons, bakers, furriers, feather-dressers, knitters, flax-dressers, &c. &c. In the words of Mr. Thakrah, "The dust largely inhaled in respiration, irritates the air tubes, produces at length organic disease of its membrane, or of the lungs themselves, and often excites the development of tubercles in con-

---

* Vide North of England Medical and Surgical Journal, No. II.

† Sur les Maladies des Artizans, p. 245.

stitutions predisposed to consumption."* These remarks refer more particularly to the filing, &c. of metals.

In the production of yarn, the cotton has to go through several processes, some of which are attended by considerable quantities of dust, and minute filaments. In the scutching and blowing department, especially where coarse and inferior cotton is used, spite of every precaution, much dust is diffused through the rooms. There are many contrivances to lessen this inconvenience, such as turning a strong current of air over the blowing machine, an aperture being made to permit the escape of flue outside—covering the machines with woodwork so as to isolate them in some degree, &c. &c. These succeed to some extent. After the first process of cleaning the cotton from foreign bodies, it is carried into the card rooms, and is here further cleaned and advanced another step towards being converted into yarn. Here also, where coarse cotton is employed, there is a quantity of dust and filaments thrown off, and it is here perhaps that most inconvenience is felt. When, however, fine cotton, of the first qualities, is used, very little dust arises from it.

The system of better ventilation and attention to cleanliness, which is spreading among the mills, have already freed many of them from the greater

* Vide Thakrah, "On the Effect of the Principal, Arts, Trades, Professions, &c. upon Health and Longevity."

portion of the atmosphere of dust, which, not many years ago, rendered it difficult to breathe in those divisions in which these first processes were carried on. A good deal remains to be done, and there is no doubt, that, as far as human invention can succeed in ridding the mills from dust and flue, this will be the case; for it is advantageous to have them as clean, &c. as possible.

Many hands are of course employed in this labour—the carding of cotton, &c., though but few, when compared to the entire complement of mill labourers: and it is here that, if evils do arise from inhalation of foreign particles, they will of course be discovered; for all the subsequent processes, are nearly free from this nuisance. In the weaving department nothing of the sort exists.

There can be no doubt, but that in many instances this dust, and these cottony particles, lead to slight irritation of the mucous membrane lining the bronchi, and produce cough and expectoration; but it may be questioned whether they induce organic disease of the substance of the lungs, or even bring on ulceration of the air passages, terminating in bronchial consumption; although a disease has been named the "spinners' phthisis. It will be remembered, that Dr. Knight, in the paper already mentioned, has very clearly demonstrated—less by post mortem examinations, than by symptoms during life certainly—that the inhaling steel particles, which are plentifully evolved during grinding, polishing,

&c., produces asthma, running into consumption. By a parity of reasoning, it has been wished to prove, that the cotton-spinner breathing an atmosphere surcharged with dust, must be a sufferer in like manner. The comparison is not fair; and the results from it must be viewed with suspicion. In the first place, this nuisance exists to any serious degree only in a few mills, and in the working a peculiar kind of material; and in the second, that this dust, where it does exist even abundantly, is of a vegetable nature chiefly, and free from the irritating qualities of metallic particles, which must in general be more or less angular, and which cannot undergo any softening, before chemical decomposition resolves them into new compounds. In these particulars, the dust floating in cotton, flax, and silk mills, is widely different from that in cutlery workshops, marble and stone grinding, &c., and most assuredly is much less prejudicial. That this dust should excite a very troublesome irritation to persons who have never been exposed to its influence, except during the few minutes spent in the mills on a casual visit, need excite no surprise: but in the parties who have become accustomed to such an atmosphere, it produces little or no inconvenience.

Dr. Kay, who advocates the opposite opinion, remarks very correctly, that "in this example as in others, is displayed that peculiar law of structure, by which it insensibly undergoes changes which enable it to endure the presence of a foreign or noxious substance, without suffering the ordinary functional derangement; and a great proportion of the operatives engaged in cotton spinning, suffer little, if at all, from the foreign particles which they inspire during twelve hours in the day. The diseases which arise from the circumstances which we have described, are chronic and subacute bronchitis." * This latter part of the quotation may be questioned as to its accuracy. In some few cases indeed, which have been brought under medical notice, a species of bronchitis, having a few trifling points of difference, when compared to the common symptoms of this affection, have been cited as proofs that they were the result of the inhalation of this dust. Symptomatology is however too uncertain a science to be worthy implicit credit, and is deserving little consideration when viewed singly, and none whatever when its dicta are opposed to the results of general observation. These cases differ so little from others occurring under circumstances where this agent could not have been applied, that it is quite obvious they are not dependent upon it; but rather that cases of common bronchitis have been rendered occasionally more obstinate in their character in consequence of these cottony particles being brought incessantly into contact with the inflamed and diseased surfaces.

---

* Vide North of England Medical and Surgical Journal, No. III.

If the inhalation of this dust was, as is asserted by these authors, followed by consumption, whether laryngeal or pulmonary, which is on all hands acknowledged to be almost uniformly fatal, there would unquestionably be found a higher rate of mortality among spinners than among weavers,—the one being exposed to this inconvenience, the other being nearly if not quite free from it. This datum, if it can be correctly ascertained, will be at once conclusive. Fortunately such is the case, as it is given in a return from the mill of Mr. Ashton, and is undoubtedly correct. "In thirteen years, during the first six of which, the number of rovers, spinners, piecers, and dressers, (*that is, those connected with the dusty part of the processes*) was 100, and during the last seven, above two hundred, only eight deaths occurred, though the same persons were, with rare exceptions, employed during the whole period. Supposing these deaths, for the sake of convenience, to have been nine,—then by ascribing three to the first six years, and six to the last seven, the mortality during the former period was one in 200, and during the latter, one in 233. The number of weavers (*that is, those unconnected with the dusty part of the processes*) during the first six years, was 200, and during the last seven, 400; and in the body of workmen forty deaths occurred in thirteen years. By ascribing thirteen of these deaths to the first six years, and twenty-seven to the last seven, the mortality during the former period was one in 92, and during the latter, one in 103."*

Thus, then, it appears, that the mortality amongst those persons who were removed from the supposed noxious agent—dust, was more than double, that which occurred among those who were exposed to its influence; and this disparity is the more remarkable, when it is considered that in the body of spinners, which exhibits such a low rate of mortality, a great number of children are included from eight to twelve years of age;—a period when many deaths occur;—whilst on the contrary, the weavers consist chiefly of young men and women from fifteen to twenty-five, a decade as little fatal as any in the course of human life.

Nothing can be more conclusive as to the question—is the atmosphere which is breathed by the factory labourer necessarily injurious to his existence, in consequence of its being loaded with the débris of cotton? Public opinion says, yes: examination into the fact says, no; though it may be admitted that the catarrhal affections to which he is so much exposed, are occasionally aggravated by this cause.

Another circumstance supposed to influence the salubrity of mill labour is, that numbers being congregated into one room;—by the process of respiration, the air being rapidly deoxygenated,

---

* Moral and Physical Condition, &c., p. 104.

is rendered unfit for the purposes of respiration.* This point has been already touched upon slightly in the chapter on Infant Labour. Wherever numbers of human beings are crowded into a limited space, the respired air is rapidly freed from its oxygen, that portion of its constituent parts which alone possesses capabilities for supporting animal life;—and its place becomes occupied by a gas produced by certain chemical changes carried on within the lungs, named carbonic acid gas,—a gas destructive to life, and incapable of supporting flame. This change in the nature of the respired air is fatal in two ways,—one by removing the oxygen, and the other, by substituting the gas just mentioned; but to be injurious in a very marked degree in either of these, the place must be confined, and the temperature pretty nearly the same with that of the surrounding atmosphere,—as elevation of temperature produces a continual current of fresh air which it is very difficult to shut out, even were this desired. Before, however, that degree of saturation with this obnoxious gas is reached, which would be immediately fatal to animal existence, the atmosphere of a room may be so far impregnated as to produce very depressing effects upon the vital activity of the system, which is attended with some peculiar consequences,—and these have been

* This deoxygenation is aided at the present day by the extensive use of gas as an illuminating agent.

well hinted at by Cabanis in conjunction with some other circumstances which have been noticed. "Dans les ateliers clos, surtout dans ceux où l'air se renouvelle avec difficulté, les forces musculaires diminuent rapidement; la reproduction de la chaleur animale languit; et les hommes de la constitution la plus robust, contractent le temperament mobile et capricieux des femmes. Ajoutez que, si la nombre des ouvriers est un peu considerable, l'alteration progressive de l'air agit d'une manière directe et pernicieuse, d'abord sur les poumons, dont le sang reçoit son caractère vital, et bientôt sur le cerveau lui-même. Ainsi donc, sans parler des emanations malfaisantes que les matières manufacturées ou celles qu'on emploi dans leur preparation exhalent souvent, presque toutes circonstances se réunissent pour rendre les ateliers également malsains au physique et au moral."

An atmosphere thus impregnated, acts indeed as a slow poison, and produces a particular impression upon the sensorium, apparently through the medium of the blood, which loses its florid colour; and when circulated through the brain in this condition, lowers all the powers of life in direct proportion with the extent of deterioration; and this is one cause of the low degree of vital energy possessed by the manufacturing popula-

---

* Rapports du Physique et du Moral de l' Homme, par P. I. G. Cabanis. Vol. II., p. 83.

tion. In former times, when the interior economy of mills was so imperfect, and the rooms were low and small, the baleful influence of this poisonous gas operated most powerfully upon those engaged in them. Latterly,—since the great improvement as to ventilation, loftiness of rooms, &c., this influence is decidedly lessened, but can never be done away with so long as the operatives are confined for twelve hours, with the short intervals of absence now allowed them.

This gas, carbonic acid, is heavier than common air,* and as fast as it is evolved sinks down to the floor, forming a stratum more or less dense according to the facilities allowed for its escape. It is thus, in some degree, removed from the influence of ventilation, as generally managed, namely by apertures or windows placed high in the walls of the room. This plan is the best and most efficacious for regulating temperature, merely as the heated air rising, according to its specific gravity, escapes freely from any opening offered it, and is replaced by a current of cold air, which rushes in to occupy its place. Not so with carbonic acid gas: it remains undulating on the floor, or falls slowly through any casual aperture, or

* A very simple experiment is sufficient to show this—"When a jar is perfectly filled with this gas, take another jar of smaller size, and place at the bottom of it a lighted taper, supported by a stand—then pour the contents of the first mentioned jar into the second as you were pouring water. The candle will be instantly extinguished as if it had been immersed in water.—*Dr. Henry's Chemistry,* Vol. I., p. 341.

flows down the stairs, but very little escapes through a common ventilator. The Grotto del Cane is a curious example of this, the depth of the stratum of gas generally being about two feet. Its existence at the bottom of wells, mines, &c. &c., is another proof of its density being greater than that of common atmospheric air, the mine or well being quite free from it except within two or three yards at the bottom. It may be questioned whether the use of cast-iron floors, now so common in mills, does not interfere with the dispersion of this gas, however advantageous they may be in other respects.

This evil is, however, undergoing a beneficial change, from the increased use of machinery requiring a less number of hands in particular rooms or mills, the gradual enlargement of these, &c. and might be still more effectually removed by directing a current of air along the floors, and leaving openings sunk through the walls.

The transitions from a heated temperature to the common day, and vice versâ, especially during certain portions of the year, are circumstances which inevitably expose the mill labourers to colds and catarrhal fevers. This effect is aided by their very light and imperfect clothing, often issuing from the mill but half-clad and without shoes or hat. This might to a great degree be remedied by the labourers themselves by a greater attention to their personal comforts.

From this cursory examination of the conditions of factory labour in general, it will be apparent, there is nothing connected with them necessarily fatal to life;—but, that some of the agencies to which the labourers are exposed exert a depressing influence upon their muscular and sensorial systems; that there is nothing in the structure of the mills, as to temperature—nothing in the material upon which they are engaged—nothing in the processes, which are, *per se*, prejudicial. On the contrary, from an extensive inquiry, it appears that the average rate of mortality has been and is diminishing since the manufactures have absorbed such an increased amount of population; but that, although life has been benefited, considered as to its duration—yet, that the standard of health has been seriously diminished. That the former is demonstrably the result of lessened vital energy, rendering the human system less susceptible of acute fatal diseases, and that the latter does not arise from factory labour simply,—but is owing, first to the use of steam, which, by requiring certain local advantages, of necessity crowds men into limited spaces, a circumstance ever unfavourable to health; secondly,—from the unintermitted attention demanded by machinery—a circumstance producing great mental and physical exhaustion—and thirdly, and this by far the most important, from the bad habits of the labourers themselves.

The diseases of the manufacturing population take their peculiar character from the causes mentioned, and are almost uniformly of a slow and very protracted kind. Few amongst the population can be said to enjoy perfect health; all are more or less ailing, and are deprived of every chance of restoration by the impossibility of removing themselves from the malarious influence which is ever around and within them. This is perhaps the hardest part of their condition, and it is a painful object to contemplate thousands of human beings toiling day after day, constantly tortured by their own diseased sensations, and driven unhappily to seek relief at a poisoned fountain, which lulls and stupifies, but only renders the mischief more deeply rooted and irremediable.

When speaking of infant labour, the situation of the factory child was dilated upon, and it was shown to be exposed to every thing which could injure its constitution. In reference to this, Mr. Green, in his evidence before a Committee of the House of Commons, remarks: "There is no disease to which children, both from the constitution of their frames, and the various unfavourable circumstances to which they are exposed, are more liable than scrofula in all its multitudinous forms. To the production of this disease, one of the most influential circumstances is, I am persuaded, breathing an impure air; and by purity of air I do not mean any thing that can be determined by chemistry: but I refer to the fact that scrofula chiefly prevails in the children of the inhabitants of densely peopled towns and crowded cities."

Scrofula may be defined a disease, or rather the name for a group of diseases, arising from imperfect or depraved nutrition, and exhibits itself in a variety of forms, all evincing that the processes of absorption and deposition are going on badly, and that the series of vessels devoted to these purposes is conveying a fluid unfitted for healthy growth, the result of imperfect elimination. This is a disease to which, *à priori*, it might be supposed the manufacturing population, in its younger branches, would be very liable,—all the conditions of their labour, localities, and modes of living, favouring the production of its peculiar diathesis. Such, indeed, observation tells is the fact: but this by no means so universally as is believed. The records of the Royal Infirmary at Manchester, containing a classification and list of diseases, are, in many points, of somewhat doubtful authenticity, no regular and systematic plan of registry being in use. Independently of surgical operations, consequent upon accidents, the majority of cases of amputation, &c. &c. are from scrofulous diseases in the joints and other parts. If a judgment were formed from this circumstance alone, it would appear that the greater number of patients seeking assistance were scrofulous. That a very considerable portion of these patients, under sixteen years of age, is affected by this disease, is perhaps true: but its effects are not marked as such; and again, vast numbers of the worst cases which fill the wards, are brought from out-districts, generally in the anticipation that some operation may be needful. Still, scrofula must be ranked as one of the principal scourges of the operative manufacturer; but it exists to an equal extent both in the one engaged in factories, and in the hand-loom weaver, who, although removed from a part of the depressing agencies of the mill-spinner or weaver, has to contend with an enemy still more destructive—actual want! In early life, rachitic affections prevail largely; but this has been spoken of in connection with infant labour.

One of the most frequent ailments, however, is derangement in the digestive organs, and this is attended by a host of troublesome symptoms, and followed by a tissue of mischiefs, which place it in advance of all other affections preying upon the manufacturing population. The diet of these people is innutritious and badly cooked. Their habits, their dram-drinking, their undrained houses and streets, are all fruitful sources of evil, and tell most decidedly upon the functions of the *primæ viæ,* altering their secretions, and producing effects in the highest degree injurious to bodily health, and reacting very powerfully upon the functions of the brain. It may be very safely stated, that one half the diseases among adults spring from this cause. The food they take is hastily swallowed, almost unmasticated; it is often coarse, and not very easily digested; it wants the stimulants of solidity and proper nutritive qualities: and, pushed into the stomach as it is, does not undergo complete chymification; hence portions are forced downwards into the intestines, and become so many sources of irritation. The constant repetition of this disturbs the healthy function of the mucous membrane lining the whole of these passages. Its secretion is changed, and it becomes exceedingly irritable, in some instances exciting the muscular coat of the intestines to frequent, irregular, and spasmodic action, giving rise to griping and cholicky pains, which are very distressing. In other examples, an opposite state of things is the consequence: the bowels are torpid, and many days, and sometimes upwards of a week, pass over and no alvine evacuation takes place. In both instances the appetite fails after a time; the sufferers lose flesh, become pale and languid, and this very often attended by hypochondriasis, which increases a hundred fold their miseries, aggravating every paroxysm of pain to a degree almost unendurable. This mental accompaniment is one of the peculiarities of indigestion; and, in the case of the labouring poor, who suffer under it, hurries them to seek relief from ardent spirits, which are a very poison when systematically or largely swallowed during the continuance of these attacks. Little wonder can be excited that, harassed by the miserable sensations to which this condition of health gives birth, the degraded labourer should have recourse to this temporary relief. He either knows not, or cares not, that it is injurious, but follows up too truly the rest of his improvidences by thinking only of to-day! Medical aid, it is true, is at hand, but to seek that he must lose his work, and a difficulty might arise as to his resuming it. He therefore struggles on till the disease takes a favourable turn, or he is so far reduced as to be incapable of going to the mill. The singularly miserable aspect presented by many of the operatives, shewing, as it were, an epitome of everything that melancholy can impress on the human face, is owing to these bowel affections. Beyond their own immediate seat of disease or derangement, they call into play a crowd of painful feelings, in all parts of the body, and are the originators of many of those anomalous diseases classed under the general term—nervous, the majority of which are dependent upon the chylopoietic viscera.

However troublesome this class of diseases may be, and it would be impossible to point out any maladies attended by a more harassing train of consequences, they are not necessarily fatal. The extent to which derangements in the digestive organs may extend, compatible with life, it is very difficult to say: not many years have elapsed since their importance became understood—a circumstance which may be explained from the change in the diet and general habits of society, rendering it not improbable that, as an extensive class of disease, they are only of recent origin. So long as the lower orders were engaged in active out-door occupations, and were supported by a simple and nutritious diet, there is no reason to suppose that they were subject to these morbid conditions of the bowels. In the higher classes of society, a tribe of diseases, having their origin in too nourishing and stimulant a diet, which produces a precisely similar train of suffering, has attracted considerable attention. It is indeed a new feature in the history of medicine, to find the two extremes of the social confederacy labouring under the same maladies, running through a similar course, and producing the same peculiar feeling of morbid irritability, intermitting with the most profound melancholy. If the situation of the pampered man of wealth, who is the victim to dyspepsia, is pitiable, how shall that of the operative be described? Language wants force to depict its horrors: unmitigated as it is, by all those foreign aids which can be procured by individuals differently circumstanced, he is condemned to labour on, a prey to bodily sufferings, and the most deplorable mental anxiety—alternately drowning his troubles in the delirium of intoxication, or standing, a ghastly and woe-worn figure, before his machine or loom.

Fever occurs in manufacturing towns, as in all other localities, but is neither more common nor more severe in its attacks. It generally assumes a low and somewhat typhoid character, and very often is attended by slight ulcerations in the bowels—a condition which has been ably described by several recent pathological writers. The manufacturing towns, from the denseness of their population, might reasonably be supposed to be peculiarly liable to epidemic febrile visitations. When it is recollected in what state the homes, streets, &c., are found, it is indeed a source of wonder that the most destructive pestilences do not periodically rage among them. Sporadic cases of fever are of course common; but looking at these towns and districts generally, they will be discovered to be exceedingly free from contagious diseases—adding another proof to those already given, that the low and depressed state of vital activity, universally characterising this population, removes it beyond their influences. Many speculations have been thrown out as to the purifying qualities of smoke, gas emanations, &c. &c., but no proofs are in existence which endow these with any importance.

Consumption has been supposed to be another disease of great extent and mortality in this population. Connected as it appears to be in some degree with a scrofulous diathesis, and exposed to so many causes of mischief in the respiratory organs, this disease is doubtless common. The imperfect clothing, the sudden changes as to temperature, &c. are powerful agents in keeping alive bronchial irritation, which occasionally runs into ulceration, or the formation of tubercles in the lungs. Still pulmonary consumption does not merit consideration, as a peculiar affection among the operative manufacturers. Neither does it appear that the deaths from this disease are increasing in any thing like an equal ratio, with the increase in the manufacturing population. Coughs and asthmatic affections are very prevalent from the same causes, but do not necessarily shorten life, or lead to organic changes beyond a slight thickening of the mucous membrane lining the air passages.

On the whole it may be said that the class of manufacturers engaged in mill labour, exhibit but few well-defined diseases; but that nearly the entire number are victims to a train of irregular morbid actions, chiefly indicated by disturbances in the functions of the digestive apparatus, with their consequent effects upon the nervous system; producing melancholy, extreme mental irritability and great exhaustion, and that few acute maladies exist amongst them: that their existence, though it is passed in one long disease, does not seem shortened; but that, on the contrary, a general improvement in the value of human life is the result of the changes which have operated on the condition of the labouring community.

CHAPTER X.

EDUCATION—RELIGION—CRIME—PAUPERISM.

Knowledge—Advantages of Knowledge—Education—Moral—Proper Education for the Labouring Man—Its Objects—Home, the best School for this Education—Intellectual Education—Sunday Schools—Their Advantages and Disadvantages—Want of Regulation—Mechanics' Institutes—Their Advantages, &c.—Infant Schools—Their Importance—The Prospect they Open to the Factory Child, &c.—Error of these Schools—Duty of Master Manufacturers—Cheap Periodicals—In what wanting—Religion—Amount and Character of Crime—Extent of Pauperism, &c. &c.

It has been said that "knowledge unemployed will preserve us from vice, for vice is but another name for ignorance; but knowledge employed is virtue." This assertion is however so far fallacious, as it assumes that vice exists alone with ignorance, and is alone compatible with it. The history of all ages and the experience of every day life are sufficient proofs that this is a position which cannot be maintained, unless by the term knowledge be understood both moral and intellectual acquirement, and these, too, so proportioned and determined, that they maintain an equal balance, a nicety of adaptation, difficult to point out, and still more difficult to reduce to settled rules.

In this extent however there can be no dispute with regard to knowledge. The ignorant man will be improvident in proportion to his ignorance, for being unaware of the evils which result to him individually, and to the good of society in general, he has no check to restrain his irregular and inordinate appetites; the ignorant man, knowing nothing of the resources open to his exertions, makes no advance in civilization or refinement; the ignorant man is much under the influence of example, and much more easily led away from the path of private duty and public propriety. The ignorant man, not understanding the principles which guide the conduct of the legislature, the particular acts of municipal bodies, the relation of master and servant, is consequently a bad servant and a turbulent citizen; the ignorant man, having no data on which to reason, necessarily forms erroneous conclusions upon most of the phenomena, whether moral or physical, which come under his notice, or he remains in stupid apathy, the plaything of accident. Education is a word applied to a series of acts having for their objects, first, the development and proper direction of man's social instincts; this may be termed the education of morals: and, secondly, the cultivation of mind, considered in reference to its peculiar attributes; this may be termed intellectual education.

The intention of educating the labouring community ought to be and no doubt is, to increase the individual happiness of every separate member; to elevate him in the scale of society, and to add to the harmony and contentment of the whole social union.

Moral education is that which, by acting upon man's feelings and principles, his natural instincts, his duties as a father, son, husband and brother, and upon his religious aspirations, has the most immediate and direct influence upon his character and happiness.

This is the education which is of the most importance to the labouring man. Circumstanced as he often is, verging upon positive want, the most demoralizing agent upon humanity, he requires all the aids which can be given to preserve him from habitual vice. He is at perpetual warfare with himself, inasmuch as his appetites and desires are more varied and more extensive than his means and opportunities can satisfy. Hence it is that morals as applied to his domestic affections and duties are of such mighty importance to his well being.

If the education which is afforded him does not aim at these ends,its proposed advantages are illusory, and its effects may be positively injurious. If the poor man's home is not rendered happier by it, he is better without it. He had better be suffered to remain in ignorance, that other classes have certain specific advantages, denied him, whilst he is shewn that he has the capabilities, the physical and political strength to aim at a summary mode of changing his condition. If his instincts and social affections are not roused and determined to their legitimate objects, his comfort is only more widely ruined; for he becomes discontented with a lot which he inherits as a birthright, and with which he might have struggled on through existence, miserable indeed in the estimation of the looker-on, but rendered more miserable a thousand times by having superadded the gnawing sense of his inferiority, which, had he been suffered tor emain in his original state of obtuse ignorance, would never have disturbed his peace.

Home ever has been and ever will be the school for moral education. It is here alone that man can develop in their full beauty those affections of the heart which are destined to be, through life, the haven to which he may retire when driven about and persecuted by the storms of fortune. It is here alone he can find refuge; it is here, that he may have about him if his condition is not supereminently wretched, feelings and emotions of the most holy and sacred influence; it is here that he may hold communion with himself; and it is here and here, alone, that he will be enabled to retain his pride of self, his personal respectability. Moral education then is that, the offices of which are to cultivate man's inherent sense of justice, to direct his religious aspirations, to make him charitable, humane, and honourable in his dealings, to bring into play his social virtues and his domestic affections, to render him a good father, son, husband, and citizen. These are the noble ends of moral instruction, and the only place in which they can be effectively developed is home; and the way in which their primary influence will be perceived, is in increasing the happiness of himself and family.

It may be safely asserted that the means hitherto in force, have not been well calculated for the attainment of the great purposes of true education. The error has lain, and it is one as yet but partially and imperfectly acknowledged, in the substitution of mere mental instruction, the acquisition of fundamental learning.

It is insufficient to teach a child to read or write; it is insufficient to teach it texts from the Bible; it is insufficient to teach it mathematics, chemistry, or history; it is insufficient to teach it a language beyond its own. These do nothing towards teaching its moral duties, and it is upon these the excellence of its character and future life must depend. As accessories, as a means to an end, and as extending the sphere of man's knowledge, these are excellent. As fitting him for improving or modifying his condition, they are excellent; as elevating him another step above the brutes around him, and as giving him a more correct estimate of himself, they are excellent;

but they should be considered as subsidiary to morals, for without them knowledge is pernicious, as it extends more widely the capabilities for being mischievous. Mere intellectual education cannot improve the moral condition of the labouring classes—it cannot render them better men or better citizens; but it can teach them their own power, and it is doing so at a period when that tuition is pregnant with danger to themselves.

So long as the homes of the factory labourer are what they are—so long as drunkenness, riot, and blasphemy—so long as all the ties of natural affection and parental subordination are unacknowledged and unfelt—so long as children are deprived of the force of beneficial example—so long as they are, both at home and abroad, exposed to viciousness of every sort—so long as depraved customs are permitted—so long will intellectual cultivation not prove the remedy for the manifold miseries incident to their present condition. Many of these have their origin in the conduct of the interested parties themselves, and many of them are the consequence of a system of labour and of payment capable of great improvement.

The benevolent founder of Sunday schools was prompted, no doubt, by feelings of the purest humanity. In the agricultural districts in which, compared with manufacturing towns, morals were in much higher perfection in ratio with intellectual acquirements, their effects were probably beneficial. The child of the agricultural labourer has a purer sense of religion than the child of the artizan born and bred in a great town; its domestic habits are formed at home, in which home, however, no means or opportunities for the acquisition of learning were afforded. To the child so circumstanced the Sunday school was an invaluable boon in some respects.

Its introduction into the large manufacturing towns, in certain points of view, was equally valuable. The incessant occupation of the factory child, entirely prevents it from acquiring the simplest rudiments of knowledge during the week-days. Shut up in the mill for fifteen hours, exhausted nature seeks repose in sleep, during the brief interval allowed it for rest.

The Sunday school to this child, therefore, gives the only possible means of gaining instruction intellectually. In another point of view it is of equal consequence to it: it prevents the desecration of the sabbath, which, were it not engaged here, would inevitably follow the want of its customary labour. It thus becomes a moral restraint, and it may perhaps admit of question whether this be not the chief benefit derived from it.

The Sunday school may be thus looked upon as the first step towards the establishment of moral feelings in the bosom of the factory child: it removes it from home—a home having nothing within or around it breathing a healthy or social atmosphere. It exhibits the day of rest in its proper light- an abstraction from habitual labour. and it conduces materially to personal cleanliness.

On the other hand, it may be questioned if means more efficacious, or at all events free from some of its evils and inconveniences, might not be devised—means that would tell more decisively on the character of home, for it is here the root of the mischief flourishes in rank luxuriance.

Sunday schools are not exclusively appropriated to the child of the manufacturing labourer; a very considerable portion of their attendants are the children of shopkeepers, petty tradesmen, and people engaged in various other occupations. Here one mischief arises from them: these, which of themselves are freed from many of the evils of the social condition of their companions, derive any thing but benefit from their association, indiscriminate as it is. It is here that great improvement may yet be made in the constitution of these schools. As they now stand, the moral good they produce on the one hand, is counterpoised by a correspondent extent of evil on the other; and when fairly estimated as to their influence on the whole community, very serious doubts may be entertained how far they have aided the moral regeneration, so much needed in these flourishing districts.

Want of due classification may be looked upon as one leading error. The selection of teachers and monitors, another. Vulgar habits and coarse minds should never have a place in the superintendence of any portion, however minute, in these schools. Brutality of behaviour, or canting sanctimoniousness, should be alike excluded.

The number of children who receive the limited education in these schools, is very considerable. This argues well on the part both of their promoters, and of those who attend them.

Mechanic institutions, intended for the instruction of "children of a larger growth," have certain advantages and disadvantages, similar to those of Sunday schools.

The mental mprovemient must again be looked upon as subordinate to the moral restraint which its members voluntarily impose upon themselves. If a man spends time, and a limited portion of his earnings in the institute, both the one and the other must be abstracted from the demands of home. But as it so generally happens, that a very small amount of the operatives actual earnings are devoted to domestic wants and enjoyments, it necessarily follows that his money must be spent, and his time engaged in the pursuit of debasing pleasures, if they can be so called; viz. in the beer-shop, the gin vault, or the political club; any time, or any money, therefore, which a wish for instruction may induce him to devote to the institution, is one step gained towards a better order of things. They are doubtless worthy of every encouragement, but it is far from desirable that they should be too exclusively made up of operatives; in fact, the greater number of masters and other respectable and superior members there are, the better it is for the true advantage of these societies.

There is one circumstance connected with these institutions, which those who augured that they would effect a complete moral revolution in the habits of the artizan must deeply deplore; and this is the gradual divergence which has been going on from their original object and intentions. Few, indeed, of their numbers are from the class they are chiefly intended to benefit.

The great moral evils which press upon the manufacturing population, are—first, those causes producing the separation of families, and the con sequent breaking up of all social ties; secondly, the early introduction of children into the mills; and thirdly, the neglect to which the infant members of the family are exposed.

To remedy this last evil—one not less in its consequences than the two others in the determination of character—a means happily has been devised, and partially adopted, which promises the best results. Infant schools have been in existence for a considerable time in some parts of the continent; but is is chiefly owing to the energy and perservance of one man that their introduction has been effected into England. Mr. Wilderspin, fortunately for the interests of the rising generation, is an enthusiast; and if enthusiasm can ever be amiable, no cause can make it more so than the one which he advocates. Without this to urge him on, he must have faltered in his course, checked and ridiculed, as he has been, and his proselytes for a time being not of a stamp to do him much real service.

To some degree he has triumphed, but very much remains undone; and were it possible that legislative enactments could be brought to bear effectually upon such a question, no statesman would deserve more nobly of his country, than he who would grant such a service to his best and most sacred interests.

Neglected, as the great majority of the infants of the factory and hand-loom labourers are by their parents throughout the day; left in charge of a mere child, or, what is still worse, a hireling nurse; badly fed, badly clothed, badly treated; their young affections blighted and seared in the very bud of their spring-time beauty—the establishment of schools, which will partially obviate these evils, is a bright spot, a break in the dark horizon, closing round their subsequent destiny, that is highly animating and encouraging to the lover of his species. It is something which, from its yet imperfect development, the imagination dwells upon with peculiar complacency, promising, as it does, to snatch the innocent victims of a pernicious state of society from a part of the miseries incident to it, and to suffer the young blossom to expand, partly at least, under the influence of a purer atmosphere.

The error into which these schools have fallen, and which, from a want of correct understanding as to the noble purposes to which they may be directed, and a corresponding want of correct knowledge of the moral evils under which their little charges labour at home, is the wish and endeavour to cultivate too much the intellect This is an error, and, still farther, it is an absurdity; for no purpose of real utility can be effected by it. The great objects of the promoters and superintendents of these infant asylums should be, to stand *in loco parentis* in all those attachments and moral associations which are inherent in young minds, and by so doing, lay the foundation for a structure which, though it may never be finished, and though it may be injured and defaced by occurrences to which it will be subjected, will, nevertheless, still shew itself at points like the beautiful proportion of a Greek temple, half buried amidst the ruins of a coarser and later period of architecture.

It is much to be wished that the master manufacturers would rouse themselves universally to forward the spread of these admirable institutions amongst their hands. They are the only parties who can do so effectually; and no mill owner could more worthily perform the duties which abstractedly he most certainly owes to those in his employ, than by establishing, or aiding in the establishing infant schools, which should embrace every child, the parents of which were manifestly unfitted for its charge, either by moral delinquency, or by being engaged in the factory.

Some of these men have, indeed, forwarded these institutions, as well as shewing in other points a disposition to assist and guide the moral conduct of their hands; their number is, however, exceedingly small. It must be hoped that the discussions which have arisen concerning the moral and physical evils attendant upon the factory system, will not be without their influence upon them As a class, they have a most extensive power of doing good, and as a class they have either by direct example or by negligence, extensive power of doing harm. Of one thing they may be assured, that by forwarding the interests and happiness of the demoralized artizan, they will in equal ratio add to their own, and that if they continue to neglect or overlook these for selfish purposes, they will repent them in bitterness and anguish.

The number of children as yet brought under this system, when all things are considered, is full as great as could be expected. Its practical advantages want explaining to the parents; probably the provident societies now establishing will do something towards this—their influence cannot be better directed.

The diffusion of correct information on the common arts of life, with facts, and moral examples, drawn from authentic and pure sources, in the shape of very cheap periodicals, will undoubtedly have their influence upon the minds of all classes. It is a vast advantage to what may be termed the accidental reader and inquirer, to have information brought to him at short intervals, and in an attractive form,—a thing aimed at by all the publications of this class.

Their diffusion, however, amongst the lower orders of the population in manufacturing towns, amounts to nothing. It is true enough that the lowness of their price brings them within the reach of most of the operatives, and it is equally true that most of them can read, but they do not purchase the weekly magazines ; the vast number sold being taken by the middle classes of society. Their place is occupied by sectarian and political tracts, which, by their exciting nature seem to harmonize better with the depraved taste existing amongst them. Very little interest or curiosity is shewn by this class of the community for the generality of subjects discussed and illustrated by the weekly periodicals. They do not come home sufficiently to their feelings and situations—they do not apply themselves sufficiently closely to their peculiar passions and wants, and they are, in consequence, neglected by them.

The literature which will succeed at present in extending itself amongst them, must come down to their moral and intellectual level—must take a much lower tone, and must be coarser and ruder in its details: facts clearly demonstrated by an examination of the works which are read and studied by them. By the neglect of this, they have as a body been left to the writings of individuals, who have written for some particular purpose, totally independent of moral cultivation, and it is desirable that the field should be now occupied by something equally homely, equally stirring, but directed to better purposes.

The results of the various efforts which have been made, and which are still making, to educate this portion of the community, are extensive and surprising, and were intellectual education alone wanting, highly satisfactory. Not many years ago, very few of the thousands that were congregated in these districts could read, and still fewer write. Now, the majority can at least read, though writing is still not common amongst them;—the very minute portion of time allowed them by the nature of their labour, must prevent any very great extension beyond this. No human being whose time, for twelve to sixteen hours per day, is occupied by exertions for procuring a livelihood, can devote himself to extend the sphere of his knowledge beyond some of its primary elements; and to do this they must be brought to his door, and be offered to him, or they will never be sought for by him. It is in vain to imagine and theorize upon that disposition of the human mind, which has had its impulses once stirred, will seek for information at the sacrifice of needful repose, or the grosser pleasures of sense. It neither is, nor will it ever be the case generally. Isolated examples may indeed be found of men struggling through the severest and most oppressive difficulties in this pursuit; but they are isolated examples—nothing upon which to found data applicable to the mass. Till the factory system is revised—till morals are established—till a code of ethics is adopted by them, no intellectual education will render them better husbands, better fathers, or better citizens; still it is a step gained—a step which, if followed by an attention to their wants by those who have the means of alleviating them, may, and it is fervently hoped will, have the most happy and beneficial result.

The little regard paid to religion by the mass of the manufacturing population, is a painful circumstance connected with their condition. A disregard for the common observances of religious worship, and the spending the Sabbath in drunkenness, being almost universal in the lowest classes. The proportion of these degraded beings, who possess no knowledge whatever of the most simple portions of Christian faith, is truly astonishing. Inquiries made amongst them have shewn, that often there is no belief in the superintending care of a Beneficent Creator—none of a state of future rewards or punishments, but that materialism has placed them in these respects upon a level with the beasts that perish. In numerous other examples, where even some knowledge was found on religious subjects, it failed in producing its salutary influence upon morals, in consequence of apathy and indifference. Thus, deprived of the most ennobling characteristic of the human mind, what wonder can be felt that it is wild waste, overgrown with noxious weeds, which choke and destroy the seeds of a better harvest, scattered as they are so thinly and so rarely over its surface. The savage, roaming through his native wilderness, bows down with reverence before the objects he has been taught to worship; and however degraded these are, they are such as his condition leads him to fear or to love, and he looks forward to the "spirit land" as his place of rest. Thus, he is religious in the only way in which his untutored mind and limited observation will permit. Thus far he is superior to that portion of the operative manufacturers which acknowledges no God—which worships no image—which regards no hereafter. The savage, indeed, from his familiarity with the operations of nature, in all their wild wonders, is impressed with her power, and yields obedience to the dictates of a whispering consciousness that there is—that there must be—a *cause* for all he sees around him. But how differently placed is the factory labourer. He knows nothing of nature—her very face is hidden and obscured from him, and he is surrounded and hemmed in by a vast circle of human inventions. He is at perpetual warfare with the world and with himself, and his bad passions are consequently in constant play. Thus combating, the pure and holy impulses of religion can find no home with him, but, like the dove hanging with trembling wing over the agitated waters of the deluge, seek a refuge in the ark of some more peaceful bosom, and leave him occupied solely by his own impure sensations.

The immoralities which stain the character of the manufacturing population, have been brought under review; these admit of no classification—no statistical inquiries can reach them. Vice, licentiousness, and depravity, which are of universal extent, can only be pointed out, and their operations watched by individuals who are led into their haunts, either by curiosity, a desire to learn in order to amend, or by professional and private duties. Crime, however, open violations of the existing laws, which subject the criminal to punishment, is more in view, and a tolerably accurate account may be given of it.

As civilization advances, and as countries be-become populous, a change takes place in the character of crime. In the early ages of society, human life was little considered; and hence, injuries to the person characterize this portion of its history. In an opposite state of things, offences against property chiefly prevail; bloodshed and murder giving way to criminal acts of minor enormity; whilst, at the same time, crimes in proportion as they lessen in enormity, become numerically greater, and of more universal prevalence.

The following table, which has been drawn up

with great care, and the results of which have been verified by examination of the authorities, very distinctly shows the varieties in the criminal character of different countries, when viewed in relation to their particular social systems.

Countries.	Popuation.	Crimes as to Person.	Crimes as to property.	Personal crimes to population 1 in	Property do. to population one in
Netherlands	6,676,000	231	935	28,904	7,140
England....	12,422,700	531	15,616	23,395	799
France......	32,000,000	1,821	17,735	17,573	1,804
Spain.......	17,732,000	3,610	2,313	3,804	5,937

In the language of the intelligent author from whose pages the foregoing table is taken:—"Crimes against property have regularly increased with the increase of commerce and manufactures, and the consequent concentration of the population into large towns. The above table shows this, too plainly to be misunderstood. Spain, the most ignorant, degraded, and uncommercial of all countries pretending to civilization, is, in respect of crime against property, *three times* less vicious than France, and more than *seven times* less vicious than England. This fact is a fearful one, and speaks volumes. England is more than twice as criminal as France, in this department of offences. We also find a striking difference between the north and south of France. *In the former, according to Mr. Charles Dupin, the increase of manufactures and trade is attested

* Forces productives et Commerciales de la France."

by the mercenary nature of the crimes there committed, as, compared with the southern districts of the same country; where offences against the person are substituted for offences against property, and are perpetrated with a black and savage atrocity, which almost baffles conception.* Ireland, again, especially the south, where manufactures have not penetrated, exhibits much violence and bloodshed, but comparatively little of dishonesty or larceny; while our own country, whose civilization we are so apt to vaunt, *far* exceeds all others in the career of mercenary crime, and has increased many years back, and is still increasing rapidly, in this painful pre-eminence in guilt."†

By a parliamentary paper, lately laid before the House, it appears, that crime, during the last twenty years, has been progressing in much greater ratio than the increase of population, but that this progression has equally characterized the agricultural and manufacturing districts. Since the late distress has so widely spread over many of the agriculturists, there is every reason to suppose criminal acts are full as frequent as amongst manufacturers. The mass of crime which prevails in these situations, originates from a variety of causes, totally independent of manufactures; and the Author just quoted from

---

* "Causes Criminelles Cèlébres, &c. Paris, 1828.

† *Enquiry*, &c. p. 24.

has mistaken the reason of the dereliction which he so correctly describes. The table which he gives to illustrate his opinions, will serve another purpose just as well. The average amount of crime as to population in eight manufacturing counties, is 1 to 840.—in the eight agricultural counties, 1 to 1,043; a disproportion much less than might have reasonably been inferred, when it is remembered how differently the two situations compared, are circumstanced. In the one, there is a thinly-scattered population—in the other, densely populous districts, resembling, in fact, a continued town. In the one an abstraction from all opportunities of committing petty crimes—in the other, a thousand incitements offer for the commission of larcenies. An examination of all towns will show, that they are placed in conditions equally as unfavourable as Manchester, in this respect. Crime is no more common in manufacturing towns and districts, than in other equally populous localities; and when the detail which has been given, as to the general immorality pervading this population is considered, astonishment should rather be excited, that crimes of greater enormity are not of more frequent occurrence. That crimes are more numerous in towns and great cities, is sufficiently well known; but with regard to England—when her agricultural divisions are found having one criminal in every 1,044 of their inhabitants; and when her manufacturers exhibit one in 840, the difference is much less than might have been predicated.

Pauperism in the class of manufacturers employed at steam machinery, is far from being of very extensive prevalence. These individuals, brought as they are into such close contact with each other, and leagued together by combinations, have also, from the force of these circumstances, formed themselves into sick-clubs, having funds for their support or assistance during illness. Most of the large mills have a particular club or union of their own; and often the spinners, weavers, &c., have each a separate one. These excellent institutions deserve every encouragement and protection, as they tend to do away with some of the evils attending upon the general improvidence of this class. That parochial relief is administered very largely in manufacturing towns and districts, is certain; but this originates and depends upon the character of the moveable population, congregated there—not upon the actual manufacturers themselves. The immigrations which are continually taking place—the surplus number of hands—and the little respect paid to parents in their old age, and when they are unable to procure a livelihood, are circumstances driving many to the poor-rates. An inspection of the past and present state of Manchester, and a comparison of these with the condition of other towns, as to the amount paid by the parish, will show very distinctly, that the major part of the population is supporting itself. Neither are the parties who claim relief from this fund, those connected with manufacture particularly;—on the contrary, but very few who depend upon these are burdensome. This arises from their unvarying occupation; a circumstance widely dissimilar from the condition of other manual labourers, who, are, in many instances, necessitated to pass a considerable portion of their time in idleness, and, of course, in poverty; and who have no resource, but to seek their subsistence from foreign aid.

CHAPTER XI.

COMBINATIONS—INFLUENCE OF MACHINERY ON LABOUR.

Labour the poor Man's Capital—Considerations on this—Its fluctuating Value—Its Rate of Exchange—Peculiar Nature of this Capital—Not Feudalism—Combination Laws—Want of Confidence between Master and Men—Combination Evils of the first—Effects—Consequences to the Men, and to particular Districts—Extent of Combinations—Their Tyranny — Their Universality — Delegates—Their Character—Seeds of disunion existing in Combination described—Effects of Turn-outs—Evil to the Men—Master less injured—Stationary Nature of their Property—Its Advantages—Union of Regulations—Consideration as to the Fate of Cotton Manufacture if entirely dependent on human Labour—Moral Influence of Combinations—Vitriol throwing—Murder, &c. &c.—Conduct towards the Masters—Combinations of Masters, &c.—Conduct of Delegates—Conduct of Masters—Importance of Confidence—Consequences of Existing State of Things—Gluts—New Hands, &c.—Benefit of mutual Regulations—Steam Engine—Application and Extent of its Powers—Proper Conduct of Masters and Men—Influence of Machinery upon Labour—Increased Production—Diminution of Wages for Quantity—Statement of this—Change of Employment—Rapid Improvement in Machinery—Superabundance of Labourers—Foreign Competition—Displacement of human Power, &c. &c,—Crisis of the Conflict of Machinery and Labour—Manufacture—Agriculture.

It has been argued by those who are favourable to combinations amongst workmen, that their labour is their capital; the fund upon which they live, and that they have the same right to turn it to the utmost advantage as any monied capitalist has to turn that to the best account. Viewed abstractedly this is true—a man's labour is his capital, it is the fund from which he derives his means of sustenance; and he has as much, and as clear a right to dispose of it in the most advantageous manner as of capital of any other sort. Labour, or the physical strength necessary for labour, to be converted into capital, must have a rateable value put upon it. It is the unrefined ore or cotton rag, the respective values of which are entirely conventional. And it is so with labour. Of itself it is nothing, by itself it is nothing—it must be stamped or moulded to bring it into a state fit for useful exchange.

Such is the mere physical capability of the working man. It would not prevent him rotting from want, and dying of inanition. Its value is given to it by the demand, and the person or community so calling it into demand, has in the first instance an obvious right to rate it as may seem at the particular juncture, its fair equivalent.

So far then the labour of the artizan is not, *per se*, of value, but its value is given to it by certain causes independent of any voluntary act of his own. The value once given, it becomes his fixed capital, and it is here the comparison commences.

In making this acknowledgment, it must be borne in mind, that if it is capital, it is of necessity liable to the same fluctuations as affect other species of capital; now, from fortuitous circumstances, doubled or trebled in value; then reduced to a like amount, or so far lower, that it ceases to be worth the possessors while to employ it.

Labour being the poor man's sole possession, his property—deserves an equal portion of legislative protection, with property of any other kind; and in return it should be content to be placed under the same restraints and regulations, which are placed for the benefit of all parties upon other kinds of property.

If in periods of great public distress, the monied men of the kingdom were permitted unrestrained by legal shackles, to advance at will the rate of interest, they would quickly absorb the greatest portion of the national wealth; if in times of manufacturing depression, the workman was permitted to raise at will the price of his labour, he would quickly absorb the wealth or means of the manufacturer. In the one instance the few would be enriched at the expence of the many; in the other, the many would be enriched at the expence of the few; but in both instances, it is quite clear that the advantages gained would be no compensation for the evils which must necessarily disorganize society.

The exchangeable value of labour must at all times depend upon many contingencies, just as the exchangeable value of the manufactured article does. The middle man as the master manufacturer may be called, placed between the labourer on the one hand, and the consumer on the other, is necessitated to accommodate both parties, and to regulate their mutual operations. If the labourer must take the arrangement into his own hands, he will naturally enough keep up the price of his own article—labour. To suppose that he will keep this well balanced, so as to enable the master manufacturer to make a profit on his capital—money; and the consumer to supply himself at a rate, advantageous, when compared with other markets, is an absurdity. He will do no such thing, he is selfish from his very circumstances and condition, and has besides that, none of those general views and extended information which are requisite for preserving a just equilibrium. He would, in the first place, ruin his master by preventing him fairly competing with his neighbours; he would ruin himself by thus lessening the immediate amount of manufacture; and he would ruin his country as a commercial one, by disabling it from carrying on its manufacturing operations in a way to rival other countries.

On the other hand the master-manufacturer, if permitted to regulate the value of labour, would undoubtedly reduce it below its just level. It is asserted that the minimum price of labour should be that which affords to the labourer a comfortable subsistence. This is unjust upon the manufacturer. The value of a man's labour, in many instances, is far inferior to what would support him. The proposition would be more tenable, were it said, that every member of a community who labours in his vocation, and is productive as far as circumstances will permit, has a claim for support upon the community at large, which, in some shape or other, derives benefit from him: but to saddle him upon a particular portion of the community, would be alike unjust and partial. Nothing should be expected from the master-manufacturer beyond this—that he should not enrich himself at the expense of his labourers; but after securing that which he has a right to do—a proper return for his outlay of capital, risk, and other contingencies, then that the surplus should be considered due to the labourers, and paid over to them in such a way as to interfere the least with their independent use of it.

Experience has long ago taught one lesson: that wherever power was possessed by man over man, it was liable to abuse, and that some regulation was essentially necessary to remedy a state of things apparently natural to him, in whatever state of civilization he is placed. It is equally for the benefit of master and men, that they should either make these regulations for themselves, or should wisely submit to any legislative enactments intended to save them from each other, and from their own partial and unjust opinions.*

* The present condition of the manufacturing population has

There is always a great difficulty in legislating for the protection of particular interests, arising chiefly from the impossibility of government acquiring correct information. Each party is alike strenuous for their separate interests, and each

---

some point of analogy with the feudal institutions prior to the reign of Henry VII., and the system of cottage building, &c. &c. threatens to make its resemblance still closer. Combinations of the men, if properly ordered, would have all the beneficial influences of the early chartered incorporations, and would produce the same effects as these did upon the prosperity and independence of the towns and bodies to which they were granted. The motives for the granting of these, and the consequences resulting from them, are pretty clearly pointed out by Adam Smith, as follows:—" In those days the sovereign of perhaps no country in Europe was able to protect through the whole extent of his dominions the weaker part of his subjects from the oppression of the great lords. Those whom the law could not protect, and who were not strong enough to defend themselves, were obliged either to have recourse to tho protection of some great lord, and in order to obtain it, to become either his slaves or vassals, or to enter into a league of mutual defence for the protection of one another. The inhabitants of cities and burghs, considered as single individuals, had no power to defend themselves, but by entering into a league of mutual defence with their neighbours. They were capable of making no contemptible resistance. The lords despised the burghers, whom they considered not only as of a different order, but as of a parcel of emancipated slaves, almost of a different species from themselves. The burghers naturally hated and feared the lords. The king, by granting them magistrates of their own, the privilege of making bye-laws for their own government, that of building walls for their own defence, and that of rendering all their inhabitants under a sort of military discipline, gave them all the means of security and independency of the barons, which it was in his power to bestow."—*Wealth of Nations*—Book III.

Dr. Robertson, in speaking of the establishments of corpo-

alike interested in perverting or distorting the facts of one another. Could they once be convinced that their interests are mutual, they would be by far the best legislators for themselves; but from the various combinations which exist, and from the exasperations of feeling which have grown up between them in consequence of these, very little hope is left that they will ever cordially unite in mitigating, on the one hand, the condition of the labourer, and, on the other, in upholding the just interests of the manufacturer.

The total want of confidence which at present marks the relations of the master-manufacturer and his hands, and the feelings of deep hatred which are too prevalent amongst them, have been brought about chiefly by unadvised combinations on both sides

It may be truly stated, that they are in organized opposition, in banded societies, for the purpose of injuring the interests of each other, from a mistaken and groundless hope that such injury would benefit themselves. It could do no such thing: and they only heighten the unavoidable misfortunes incident upon their separate states, by a course of proceeding at variance with every

---

rations, says—" The good effects of this new institution were immediately felt; and its influence on government, as well as manners, was no less extensive than salutary. A great body of the people was released from servitude, and from all the arbitrary and grievous impositions to which that wretched condition had subjected them.

thing just and charitable, and which, by its demoralizing agency, is rapidly unfitting them from ever regaining a position, with reference to each other, which is alone compatible with their best interests.

Combination is justifiable only when a disposition is plainly shewn to take certain advantages which may be more or less injurious to either party. On the one hand, the master may wish and endeavour to exact more work from his hands without increasing their wages, and thus add to his profits at the expense of their labour. Here he may find himself thwarted in his purpose. His men scatter themselves amongst his neighbours, or refuse to work, knowing that he cannot afford to let his machinery and stock in trade remain idle. Under these circumstances, if his neighbours do not embrace his cause, he is compelled to take back his hands upon their own terms, having possibly to make certain sacrifices, as to mill regulations, as a propitiation. The passive combination of the men here gains them their object; and, in this particular instance, perhaps the conduct of the individual master might be wrong. But this triumph by no means leaves them in the same relative situation which they held before. Their natural order is to some extent reversed, and the men have learnt a dangerous power. Farther still, the character of the master is lowered in their estimation; they have no faith in him, and are eternally jangling and disputing upon points of discipline. On the other hand, this power having once been acknowledged, the men, in their turn, become more unreasonable, and during a run of trade, or seizing upon some other favourable juncture, demand either a lessening of the period of labour, or an increased rate of payment for that which they already go through. The master demurs; his hands strike, and he finds himself on the very verge of ruin in consequence. He holds out if he can; and the men, having their own sufferings to contend with, in the end return at their old prices, or an increase so small, as to be a straw in the balance compared to what they have lost during their wilful idleness.

It is quite obvious that occurrences of this nature, so detrimental to the interests of the men on the one hand, and the masters on the other, must lead to the adoption of some measures having for their intention the equalization or protection of both against the caprice, avariciousness, and unreasonable and untimely demands. Unfortunately, each party made their own arrangements. The men under the belief that they were all powerful, and the masters in self-defence, with the farther understanding that they would assist each other. On both sides funds were collected, delegates and secretaries appointed, and labour and monied capital came into direct collision.

A history of the privations borne by the workmen in some of these insane contests, would present an awful picture of human suffering, and a picture not the less awful of rapid demoralization. So far have these gone sometimes, as to threaten the ruin of an entire and flourishing town or district, and have involved in it not only the interests of the two conflicting parties, but the race of shop-keepers and others dependent upon them for their support. The most extensive emigrations have taken place, poor-rates have been doubled, and society disorganised.*

These struggles have uniformly been most disastrous to the men, and must ever be so. It is in vain that in their rage worked up into madness by heartless demagogues, by hunger, by the sight of their famishing children, they have taken the law into their own hands, and dreadful proofs have they given how unfit were they to wield it for their own benefit. Incendiarism, machine breaking, assassination,† vitriol throwing,‡ acts of diabo-

---

* The effects arising from combinations are almost always injurious to the parties themselves. As the injury done to the men and their families is almost always more serious than that which effects their employers, it is of the utmost importance to the comfort and happiness of the former class that they should themselves entertain sound views upon this question."—*Babbage*, p. 296.

† The deliberate assasination of Thomas Ashton, son of Mr. S. Ashton, one of the principal cotton manufacturers in the neighbourhood of Manchester, during the strike in 1831-32, was an act of the most atrocious villainy. Returning from the mill early in the evening through a bye-road, he was shot through the chest within a very few yards both of the mill and his father's house—a victim less deserving his untimely fate it

lical outrage, all have been perpetrated for intimidation or revenge; but in all cases with the like result, or when partial success has attended them, it has been but temporary.

The extent to which combinations exist amongst workmen is only fully shewn when a general strike in a particular branch of trade takes place. Then they are seen ramifying in every direction, embracing all trades alike, each having their separate rules and code of laws, but all uniting

---

would have been difficult to have selected—for he was distinguished by his general kindness to the men, and endeared by his amiable qualities to his friends. The crime would seem to have been perpetrated by three people in company, who had been seen lurking about during the early part of the evening. Large rewards were offered both by government and his connexions, but have hitherto had no result, and it seems probable that its perpetrators will escape the punishment so richly deserved by their act—save the hell of their own consciences—or that the strange retributive justice, which has marked similar atrocities, will in course of time bring them to light.

‡ The crime of vitriol throwing is a novel feature in the annals of the country. It consists of putting into a wide necked bottle a quantity of sulphuric acid—oil of vitriol as it is commonly called—and throwing this upon the person of the obnoxious individual, being either directed to the face or dress merely, or of throwing a quantity upon any work offensive to the party. The caustic nature of this fluid renders it a formidable weapon when it is applied to any exposed part of the body, and in several instances loss of sight has resulted from it, and in a vast number of others very great suffering has resulted from its application. It is immediately destructive to the texture of cloth from its corrosive qualities. This demoniacal proceeding was exceedingly prevalent during the turn out of 1830-1, many of the masters not daring to stir out during the evening.

in one point, to support the operative, when he either voluntarily abstracts himself from employ, or is driven out by some new demand on the part of his master. Each trade has a sort of corporate board for the management of its funds, the protection so called of its particular interests, and this board is paid for its services out of a specific allowance made by every workman who is a member of the union or combination. The sums thus abstracted from the pockets of the deluded artizan, have been all very considerable, the regulations being compulsory in the extreme.

No workman is allowed to act according to the dictates of his own feelings—he is compelled to become a member, or subject himself to a course of annoyances and injuries which have repeatedly ended in death. This arbitrary and tyrannical assumption of power, is one of the greatest evils attending upon the system of combinations amongst the operatives. However well-disposed the industrious and economical workman may be, he is placed upon a level with the most profligate and idle, who are in general the stirrers up of these strikes; he is condemned against his own judgment to abstain from working at a price, low perhaps, and barely sufficient, but still enough for his wants; to be satisfied with two or four shillings per week from the union fund for the support of his family, in the place of earning amongst them twenty or twenty-five shillings; to remain in idleness for weeks in succession, to the utter ruin of his habits—and is deprived of all stimulus to be a good and industrious citizen, by the certainty that he is liable to be turned out of employment from causes over which he has no controul, and which not unfrequently he cannot comprehend.

When it is borne in mind how great a proportion of the labourers employed in factories consists of females and children, the evil presents itself in a still more striking point of view. In nineteen cases out of twenty these can have nothing to do with the originating the turn out, which is equally, however, operative upon them, a regular compliment and series of hands being necessary for the working of a mill. They are thus subjected to starvation and idleness, both exercising a most powerful influence upon their moral and social character—an influence of the most debasing quality.

The operations of manufacturing yarn from the raw material, it is true, is distinct from the manufacturing it into cloth—so far the spinner and the weaver are completely separated; they are, however often carried on simultaneously in the same mill, the same steam-engine serving for both purposes. Occasions have occurred when the spinners have turned out leaving the weavers in employ, and vice versâ the weavers have turned out leaving the spinners in employ, each supposing itself to have separate causes for complaint or satisfaction. The great turn outs have, however, involved both, or when they have been partial, the injury, though less, has still been felt by the whole—the members of the same family or household being indiscriminately composed of spinners, weavers, and their dependants.

It has repeatedly happened that the disputes which have ended in a general turn-out have had reference only to a very small portion of the hands—spinners of a particular class for example, such as coarse or fine yarns in the demand for which some change has come on, which may have necessitated the master to reduce the wages paid for its manufacture, but so complete and determinate has been the organization amongst the whole union, that thousands have deserted their occupation and submitted to every sort of suffering incident to the deepest poverty.

Many writers, at various times when the public mind has been disturbed by these disastrous divisions, have stated that the funds of the unions have to a very considerable extent shielded the turn-outs from sufferings. When a turn-out is but partial—that is, when the hands of a particular mill, or of a limited portion of one of the great division of manufacturers, from causes peculiar to themselves, have struck, this statement is to some degree true, but in the general turn-outs, when from ten to eighteen thousand men, women, and children have to receive assistance from these funds, they are but a drop of water in the bucket, even though assisted in the most extensive way by the shopkeepers, and by advances from every other trade.

The appointment of delegates, individuals chosen from amongst the workmen, and invested with power to arrange disputes and to manage their funds, is a practice deserving the utmost reprobation. By withdrawing the body of the workmen from confidential intercourse with their masters, the interposition of these middlemen, like the Tribunes of ancient Rome, increase the evil a thousand fold. Living as they do upon a liberal allowance from the wages of the starving operative, their motives are open to suspicions of the most injurious tendency, and the character of most of these does anything but remove them. Selected at moments of party excitement, of rancorous feelings, of raised expectations—the brawler, the factious man, the specious scoundrel, have too often become the dictators of the misguided people, and to maintain their own evil pre-eminence they have spared no pains to distort and garble facts, to blacken and destroy the reputation of the masters, and to keep open the breach from which they alone could derive advantage. For a considerable period subsequent to the first organization of the unions, the men were misled by the numbers which formed them—and their calculations based on numerical strength deluded them into the belief that they were all powerful. This hallucination, for it deserves no other term, was carefully fostered by these delegates, who finding themselves elevated into petty kings, wielded their authority with as much complacency and despotism as other potentates. For a time it is possible, from the narrow and selfish view they took of matters, they were equally deceived by the universality of the combinations, and indulged the hope that they should be able to crush the masters, and make them subservient to their dictates.

These combinations had, however, within themselves the seeds of discord and disunion—the elements of dissolution. Could they have been brought to conform to fixed and general laws, and to act simultaneously, they would indeed have been irresistible. Great bodies of men, dependant although they may be upon some one branch of national industry, have nevertheless separate and well defined interests. During the first burst of enthusiasm, brought about by what they think emancipation from thraldom, these may be forgotten or thrown aside, and all may join heart and hand in the promotion of what they suppose the cause. If at this time any man of commanding talents, and great energies, were to arise amongst them, he might become the agent for prolonging this state of things. Men gifted with the requisite talents and influences have, however, as yet, fortunately for the welfare and commercial existence of the kingdom, never arisen. Trumpery demagogues have indeed pushed themselves into short-lived authority and leading; but none have been found, but who have soon displayed their entire inability to rule or direct the storm, which merely bore them on its surface. No sooner were the effects of the vast drain upon their money resources felt—no sooner the first excitement passed away—than the natural interests and importance of each separate town or body were again brought into play, and want of general cordiality and unanimity of acting soon rendered the great combinations partially inoperative.

The scenes which were the result of these combinations of labour against capital and due subordination, all who have witnessed will join in condemning as tending to the destruction of social order, and the security of life and property; and though the arm of the executive may have fallen heavily and indiscrinately upon the disturbers of the public peace, no man having a correct knowledge of the results which would have necessarily emanated from them, can deplore or condemn it.

No estimate, at all approaching to accuracy, can be made by persons living in districts or towns which have never been visited by one of these turnouts, of the social disorganization which atends upon them. In 1805-6, when power-looms were first introduced, the total suspension of trade, and the turbulent and distracted state of the handloom weavers, with the consequent violations of rights of property, in the destruction of machinery, are still well remembered. But this display of hostility by no means interfered with the rapid spread of the power-loom; on the contrary, it only afforded a more decided stimulus to it. The manifestation which had been made, previous to this epoch, of the difficulty of keeping in due subordination human power, made steam doubly welcome to the manufacturers, by placing under their control an agent, which, their sagacity clearly foresaw, would, in the course of time, render them to a very great degree independent of it. And events have justified their prognostications; but they failed then, and have since failed in convincing the men, that they were pursuing a line of conduct, which, if permitted to proceed to extremities, must ruin them by ruining the manufacturers, and by driving away capital, for ever deprive themselves of the benefits invariably resulting from its employ, and in shewing that their opposition to improvements in machinery, which were gradually lowering their wages, only tended still farther to accelerate their progress.

One great evil which has ever attended both partial and general strikes that have continued for any length of time—and some of these have extended over six months—is, that the means of the turn-out have been quite exhausted, long before he again got to work. The pittance he received from the Union did but little towards his support, and hence he was necessitated to get into debt, and pay a most extortionate price for the common necessaries of life. These engagements were, of course, to the retail dealers in provisions, who were themselves not unfrequently ruined by a system of credit, far too extensive for their limited capital. In all cases, however, the long arrear had to be paid up whenever work was obtained, and this absorbed the greater portion of their wages for many months,—thus prolonging the distress which had already well nigh overwhelmed them. This has been one great cause of the squalid wretchedness so generally found in the houses of the operatives.

It is obvious, from the preceding details, that combinations of workmen against masters—of labour against capital—have uniformly, in the manufacturing districts, been injurious to the men to a much greater extent than to the master. His capital, it is true, was for the time injured, but he had means of supporting himself during these struggles. The nature of his investment prevented him from carrying away his capital and enterprise to other localities, and other ways of employing them; and in this has consisted one great cause of the stability of manufacturers. Building and machinery could not be removed nor converted into portable property, without a ruinous sacrifice. Often, indeed, the manufacturer has been heard to deplore this as his greatest hardship, which was in point of fact his greatest advantage, for it prevented his abstraction, and by forcing him to submit to a temporary inconvenience, he was sure, in a short time, to be right again. The same fact—and it is one which strongly illustrates the almost paradoxical proposition, that men rarely know what is their real advantage—induced the workmen to believe, that thus having the property of their master fast in its present position, they could the more easily make him compromise himself. It had, in point of fact, exactly the opposite effect in both cases. On the side of the master he was chained down, and thus compelled to wait for a revulsion in the popular mind; on that of the workmen, in spite of all the precautions they could adopt, new hands were invariably introduced, or means taken to secure their introduction, when the storm was blown over; added to which, the mill-owner, thus deserted by his men, turned his attention and efforts to the improvement of his machinery, and acquired a practical knowledge of its operations, and capabilities for alteration, which, under a different state of things, he might possibly never have acquired.*

The general combinations, which include nearly the whole of the manufacturing population, and which are so mischievous in their effects, have numerous minor and subsidiary co-operative unions. The spinners, weavers, piecers, finishers, &c. &c. each combine in their own class, and have each distinct rules and regulations for their direction; subordinate, however, to the general confederations into which their distinct interests merge. The same evils in a minor degree attend

---

Vide Babbage, p. 298—Manufacture of Gun Barrels, &c.

upon these; the divisions amongst them being founded chiefly with regard to the amount of wages received, and the differences in the amounts of contributions paid into the general fund. They are, however, quite distinct in their separate corporate capacities, acting quite independently of one another, and assimilating only in being parties in the general combination.

Independently of associations for the fancied protection of the general and separate interests of the several classes which compose the manufacturers, and which relate to general questions connected with their interests, every mill has its peculiar arrangements. These are regulations as to hours of work, fines for imperfect work, regulations as to the engine, &c. &c., and are commonly made in conjunction with the master, and are indeed necessary for the interests of both, as they serve as mutual checks, and as restrictions upon the conduct of the overlookers, engine men, and are thus found mutually advantageous.

In looking back at the history of the cotton manufacture, and in examining the rise and growth of combinations amongst workmen, it may be asserted, that but for the application of steam, it must either have been destroyed by those who should most have fostered it, or that it would have been so restricted in its operations, and so burdened with expensive details, that it would never have progressed so as to become the staple trade of the country, and thus, too, quite independent of the additional power gained by it. To what extent hand manufacture might have gone, it is not easy even to guess. But is quite certain that it might have gone vastly farther than it has ever yet been pushed. If all the individuals now engaged in this manufacture, were devoted to hand labour, an immense amount of production would be the result, though infitely inferior to that now existing. But there is no reason why double the amount of hands should not be engaged in it, or treble, or any amount that would be sufficient to throw off as much work as would equal the present joint production of steam and hand labour.

Such a mass of labourers, brought into intimate communion with each other would, however, be utterly unmanageable. Trade would be liable to fluctuations which must ruin it, and its entire control would be vested in the workmen. If combinations have been ruinous to some masters, and injurious to trade at some junctures with the aid of steam, in some degree an independent power, what would have been their effect if it must have entirely submitted to their caprice?

The moral influence of combinations upon the character of the manufacturing population, has been exceedingly pernicious. It has placed a barrier between the master and the hands which can never be removed, but with the utmost difficulty; for neither dare trust the other. Suspicion has usurped the place of confidence; an utter alienation of all friendly feelings—mutual fear—hatred, and a system of espionage totally subversive of every thing honourable in their intercourse.

Nothing can more strongly mark the demoralization brought about by the agency of turn-outs, the result of combination, than the acts which have been committed by the men, and openly boasted of. Cutting away cloth from the looms of those better disposed individuals who preferred working for the support of their families, to starving in obedience to a fiat to strike—a system of inquisitorial visitation ruinous to neighbourly kindness—the darker crime of vitriol-throwing—more cowardly and treacherous than the stab of the Italian bravo—the waylaying and abusing obnoxious individuals—and the stain which must ever rest upon them, indelibly marked, and branding them with infamy—that of murder.

The demonstrations of vengeance, carried into effect in some instances have at times forced the masters to arm themselves—to garrison their houses and mills, and have placed a whole district in a state of siege, with all the disorganization of social ties incident to civil warfare. No man was safe—no family secure from midnight disturbance; shots were fired into the rooms where it was believed the master had his resting place. By day he had to use every precaution to avoid falling into the hands of an infuriated mob—his family reviled—loaded with the most opprobrious epithets—hooted and hissed wherever opportunity offered;

no wonder that feelings of bitterness were roused against those who thus wantonly violated all the forms of decency and justice.

Another moral evil of equal magnitude to the labourer, resulting from combinations, is, that it renders himself and family improvident, and this is the very bane of all his comfort. Whatever the cause may be which renders a man's labour of doubtful continuance, or which occasionally, and at uncertain intervals, throws him out of employ, it invariably injures his industrial character. The same remark has been used before, when speaking of the hand-loom weavers, whose occupation was irregular and fluctuating. It was very truly observed by Mr. Galloway, in his evidence, that "When employers are competent to shew their men that their business is steady and certain, and that when men find that they are likely to have permanent employment, they have always better habits and more settled notions, which will make them better men and better workmen, and will produce great benefits to all who are interested in their employment." In these respects the combinations and "strikes" have been ruinous to a degree. A consciousness that there is this uncertainty with regard to the continuance of his work, removes from the labourer all stimulus to work diligently, and earn for himself an honest reputation. He feels that he is no free agent in the matter, and loses, in consequence, much of that personal respect for himself without which he possesses but half his moral stamina.

The repeated evils brought upon the masters by the combinations of their hands, made it imperatively necessary that they should be met in a similar way. No master could cope single-handed with a combination of his own men, who were leagued together to prevent any sufficient supply of new hands, and were liberally supported by their co-mates. The interests of the masters having become more generally understood, and that of one identified with the rest, they in turn refused to employ any operative who made himself conspicuous in the combinations, and gradually feeling their own strength, dismissed all those who could be detected as belonging to the obnoxious unions. Not satisfied with this, however, they refused to engage a workman who had been dismissed or had voluntarily withdrawn himself from the service of another master provided he did not bring with him a character. This once gained, the masters had the destiny of each operative in their keeping. If he disbehaved himself he was straightway dismissed, and found himself, unless under very peculiar circumstances, utterly excluded from all chance of getting occupation in that town or district, and had no little difficulty in procuring it even if he removed. For a time he received a pittance from the joint funds, which spite of all the vigilance exercised for its suppression, still existed.

So long as masters confine themselves to the legitimate objects which should govern their unions; viz. protection against the unfounded and unreasonable demands of their workmen—the cultivation of their own undoubted authority, with the interior regulation of their mills—their efforts will be hailed and cordially seconded by every man who has a knowledge of the miseries incident to the want of these, and who looks forward to a regeneration in the moral and social condition of the operatives. It is much to be feared, however, that these combinations may not end here. The terrible lessons which have been taught the masters will lead them to exert every device, to strain every nerve, in order to annihilate the influence of the men beyond simple and subordinate agents to their tractable and gigantic servant the steam-engine; and in doing this, both the interests of the men, and that of the public, might very probably suffer.

It is very true, notwithstanding these changes, and notwithstanding the diminished influence of the general combinations, partial out-breaks, directed against a particular master, still occur, shewing the smouldering fire which it is so desirable to extinguish. In these cases it however generally happens, that the causes in dispute are entirely local and confined to a particular order of workmen, often indeed arising from trivial alterarations introduced by the owner in some particular quality of work or mill regulation. The old spirit exists in this respect of striving by every means of intimidation, and even resorting to acts of personal violence, to deter new men from being engaged. Even at the moment this page is writing, a scene of this nature is going on within fifty yards of the author's study window, accompanied with all the confusion naturally incident to such outrage.

If experience could have taught the men wisdom, they would long ago have learnt that riot could serve no good purpose, and that now the masters are banded together, and know their interests to depend upon the exclusion of turbulence from their mills, it must be fatal to any party of them who indulge in it. They cannot prevent the introduction of new hands, they cannot prevent improvements in machinery; and though their wages must suffer, and though their labour must go on increasing to acquire even the wages they now earn, there is no human power to interfere with or thwart it.

The moral effects of combinations amongst the masters are, the keeping alive those suspicions which have so long been the bane both to the happiness of the labourer and to their own. Ever disposed to view all attempts at change as innovations upon what he erroneously conceives to be a matter under his own control, the workman murmurs and finally rebels, if he finds that these changes are tending to increase his hours of labour without an adequate compensation. As he finds, however, that his rebellion has ceased to be beneficial, that his master assumes a higher tone, and, spite of all his efforts, contrives to carry on the operations of his mill partly by the introduction of new hands, and partly by improving his machinery, he will surely discover that he had better, in place of becoming a turn-out, endeavour, by exertion, by economy, by domestic habits, by acquiring feelings of providence and anxiety for something beyond to-day, remedy, as far as is in his power, the change in his circumstances. If masters will combine steadily, and aim at inculcating subordination, respect for proper authority, and to these add their anxious endeavours to effect these desirable alterations in the present characters of their workmen, which can alone render them permanently quiet, they will deserve well of their country. If they succeed in establishing their own power upon the wreck of the labourer's independence and morality, they will have nothing to congratulate themselves upon, and must ever live on the surface of a volcano, which at any moment may be roused into destructive action.

There are questionless many of the great cotton-spinners who are ardently desirous to see an amendment in the moral and social condition of their hands, and who take every means in their power to effect this. Hitherto these efforts have been attended with but little success, and some of them have raised the cry, that it is useless to attempt to influence the conduct of these men ; that they meet with nothing but the basest ingratitude, and that the best thing they can do is to reduce them as rapidly as possible to mere automata, and deprive them of every inducement to exercise their will, by grinding them down by labour and starvation. That they have met with repulses sufficiently disgusting and disheartening is very probable, but they suffer their passions to run away with their judgments when they condemn the men for expressions of hatred and distrust which are explicable, and if not excusable, may be so far palliated by circumstances and considerations as no longer to brand upon them the impress of stupid and brutal ingratitude.

Years must elapse before entire confidence can be restored between the opposing parties, if both do not cordially join in striving to forget and forgive injuries and insults—innumerable, and bitter to the memory. But upon the masters, as men of education, men of extended and enlightened view, the principal part of the task devolves very naturally and properly. Let them combine strictly, let the law give them its utmost protection,—but let those powers and combinations be tempered with justice and mercy. They have the social, moral, and domestic condition of the men intimately under their influence and command—let their example, in the first place, afford them a lesson of order, decency, proper economy, and religion; and secondly, let them remember that excess of labour partially unfits them for the display of those virtues which can alone give them any chance of happiness or comfort.

It is morally impossible that the condition of the great bulk of the labouring community can be ever permanently benefited so long as they permit themselves to be led away by designing and factious individuals, who live upon their passions and ignorance. Till a restoration of confidence between them and their employers they will be liable to this misleading. Whenever any question having reference to the interests of either party is under discussion, these agitators immediately stir themselves to effect a breach between the employers and the employed, and it very frequently happens, that the first intimation that a master receives of his men interesting themselves on any subject, is the sight of a petition containing the signatures of all of them. This display of want of confidence on the part of the men, and the secrecy and sometimes little understood way in which they receive information, renders the masters jealous and suspicious, and spite of their prudence, they shew their feelings pretty plainly. Hence it is that so little progress is as yet made towards a proper understanding between them, whilst the actions of each are scrutinized closely; when, on the one hand, it should be ready obedience, and, on the other, confidence and affability.

Intellectual education, by partially expanding the views of the operative, has as yet done little towards improving his correct judgment, and enabling him to detect the mischief that may lurk beneath the disguise of apparent sympathy when painted in glowing colours. Opportunity has been taken of this by their delegates, &c. &c., and they still succeed to a certain extent in some situations in keeping up their injurious influence, driven as they have deservedly been from the mills of those masters who are anxious to preserve order, and establish a good understanding between themselves and their men: they have either been compelled to work for the inferior employers at a less wage, or have sought some other business ostensibly, but still trusting to derive subsistence from the occasional ebullitions on the part of the prejudiced and irritable workmen, and in some instances they are successful. The bad passions of the men are easily roused—they have no faith in the justice of their masters, and though a state of things is gradually encompassing them which must force them to succumb, or entirely ruin themselves, no sooner does some improvement in machinery of extensive operation lower the value of labour, and render certain changes necessary either in wages or hours of work, than bickerings and heartburnings arise, which however, for some time have subsided peaceably. The example shown to the men during 1830 and 1831, when the masters one and all stood out to resist certain demands, did much towards breaking the spirit of insubordination, and if they remain firm and decided, and mutually assist each other, they will eventually succeed on putting an end to combinations and their concomitant crimes.

The establishment of schools by the master, the taking a personal and positive interest in the welfare of those immediately under them, the shewing them that they are sincerely anxious to assist, and pointing out the way in which their wages may be made amply sufficient to cover all their wants, will in time produce their effects, in all out districts—in the towns the process of regeneration must be slower. The men are much less under the influence of the masters, their hands making only a part of a great and demoralised population.

The improvement which has taken place in the great body of masters within the last few years, is perhaps the best guarantee for improvement amongst the men. The immense amount of capital invested in the cotton trade which has been calculated at not much less than fifteen millions sterling, will prevent them unnecessarily doing anything which can involve in danger the returns for this outlay. They are too, fully aware, that their locality is determined, and that, however, turn-outs may plague and harass them that they are fixed and stationary. A general turn-out even for a limited period, is attended with enormous pecuniary sacrifices, but falling as it does upon great masses of capital, or upon high reputation and credit it is comparatively harmless. Any lengthened idleness, however, in this capital is ruinous in a high degree, both to the trade itself and to the masters. This is one of the most powerful reasons for inducing them tc cultivate the confidence of their men, and for joining in an union, which will enable them to protect themselves, by refusing to yield to clamour, and in arranging certain amounts of wages to be paid by them as a fixed rule.

The higher the character of the master ascends, and the greater his wealth, and the gradual clearing away from amongst their body mere men of straw, who could submit to no rules on account of their necessities, and the approximation to one general standard will enable them in time to lay down such rates of wages as will be universally applicable; some time must necessarily elapse before this can take place. The great manufacturers, who also invariably possess the best machinery, can afford to make goods at a something less price than the secondary men. If a strict combination were therefore to take place amongst them, and they were determined to sell as low as possible and keep their wages down, they would speedily ruin the inferior manufacturer, who having a limited capital, second rate machinery and an absolute necessity for quick returns, could not meet them in the market. A tacit agreement seems to exist, by which the operative and manufacturer at present benefit at the expense of the public, and this enables the inferior man to make a profit under all his disadvantages. The outcry as to the ruin of the trade is a positive absurdity—it may be confidently asserted that those manufactories which are extensive and upon the most approved arrangement as to machinery are making immense returns, and the proprietors could afford to work them, and sell at a very considerable diminished price.

If the workmen could be brought to understand some of these truths, a better chance would be given for a union taking place between them and the masters, and both would succeed in keeping up a remunerating price for their separate capitals; whereas the men, on the one hand, by injuring the property of their masters, compel them to reduce their wages as a compensation, or to demand such excellence of work, and to cut off so much for fines, that it amounts to the same thing as reduction; whilst on the other hand the master, finding his hands thus turbulent and unruly, turns all his energies to improvements in his machinery, and endeavours, by throwing off an increased quantity of work, to make amends for the incertitude in his labourers. This often produces gluts, which injure, again, both masters and men. Another advantage would be gained by a mutual understanding: during turn-outs great numbers of new hands come into the town or district where it exists, generally hand-loom weavers, or operatives from other classes of manufacturers, or individuals from the mining districts; and, rather than have their mills remaining totally idle, the masters engage these, spite of the violence and resistance of the old hands: but then they are ignorant of the details of spinning or weaving; much has to be taught them; a great deal of work is spoiled, and much loss sustained for a time. The men finding their places partially occupied, and their obstinacy and turbulence yielding before continued privations, gladly return at the old prices: many of necessity cannot be taken in, as it would be a manifest injustice to discharge the hands they have, even though inferior in capability for producing good work. Thus they are compelled to seek work somewhere else, or lower their wages, and have been instrumental in introducing a greater number of hands than can find present employ, and hence the foundation for another reduction of wages before any very long period.

The most important benefit which the men would derive from coalescing with their masters, and agreeing to some certain standard rules for their mutual regulation, is that both would very soon think each other trustworthy, and those feelings of hostility and bitterness which now exist between them, and which are kept up by combinations—delegates—secretaries—would be done away with. It will be in vain for the men to expect to do this, so long as they submit to the dictates of parties interested in keeping up mischief between them and their masters. The moral revolution which this would at once produce, would be an excellent basis for rearing a superstructure of social arrangements, which might snatch the men and their families from their present degradation. One thing is quite certain, that if they do not adopt some plan of this nature, they will miss the opportunity, and will condemn themselves to a life of servitude to an iron master, who is already more than threatening them. Let them remember that already the steam-engine, though applied to the same purposes as human labour for so very few years, performs as much work, in simple power, as two millions and a half of human beings! Let them farther remember, that each steam-loom is nearly four times as effective as the hand-loom, and that improvements are hourly taking place in its applicability—giving it endowments—approximating it with the most delicate operations of the human hand; and let them remember, also, that it never tires; that to it eight, ten, twelve, fourteen, or twenty hours are alike! To endeavour to arrest its progress would be madness: they cannot turn back the stream of events—the onward current of the age—their efforts would be equally impotent and ruinous. They may, however, compete with it on more equal ground if they choose, and may prevent the accelerations of its career by working steadily, orderly, and systematically. Every effort which they have made to retard or destroy the progress of machinery, has only hurried on the march of improvement; and if they persevere in such a course of proceedings, they will become its victims, bound hand and foot, and resembling, in their condition, the serfs—the *glebæ adscripti* of a former period. Their attempts to break up the social confederacy by violence and outrage, and thus bring on such internal disorder as would deprive the nation of its manufacturing pre-eminence, must and will be suppressed by the strong arm of the law.

Let the masters, then, combine—let them elevate their own character—let them become great and wealthy; for greatness and wealth, when joined to correct knowledge of the wants of those around them, are the best masters. Let them, by every means in their power, aid the moral and social regeneration of their men; and let the men meet them frankly and fairly; let them be assisted to shake off habits which destroy them physically and morally; let them cultivate home; let them become good husbands and good wives, and they will, in a single generation, produce offspring who will, in their turn, inherit their good qualities. Let them discountenance agitation, combination, and political quacks. Let them become a sober and orderly race; let them, if they will, appeal to the legislature for enactments to regulate their labour, if excessive— and by so doing they will be their own best friends. With economy and foresight, the means they possess are sufficient to supply their wants, and the requisites of their stations; and were these doubled, without economy, they would be no real gainers. Let them bear in mind that all the members of their family have a resource in the mill; and though the individual earnings may be in some instances small, yet that, collectively, they are considerable, Let families reunite; let them eschew the gin and beer-shops,—and they will yet be a happy and contented population. But if they do not, they may depend upon this,—that they will be ground down to the earth, and present the most humiliating and miserable picture of humanity which the world has ever yet seen—save only the slave-gang, and the Indian miners, under the deadly yoke of the early Spaniards.

It has been truly observed by Mr. M'Culloch, "that the operatives are in great measure the architects of their own fortunes; for what others can do for them is trifling indeed, compared with what they can do for themselves. That they are infinitely more interested in the preservation of public tranquillity than any other class of society; that *mechanical inventions and discoveries are supremely advantageous to them;* and that their real interests can only be efficiently protected by displaying greater prudence and forethought."*

The influence of machinery upon human labour, has some points which connects the question closely with that of combinations amongst the masters. The remark in the foregoing quotation from M'Culloch, "that mechanical inventions and discoveries are supremely advantageous to them," *i. e.* the manufacturing population, is far too ge-

---

* Edinburgh Review, No. XCI.

neral and sweeping, and an attentive examination of the subject would have sufficed to satisfy the author that he was suffering imperfect data to mislead his judgment, and was hazarding an assertion unsustained by facts, whether of observation or reasoning.

"One very important inquiry," says Mr. Babbage, "which this subject presents, is the question, Whether it is more for the interest of the working classes that improved machinery should be so perfect as to defy the competition of hand-labour, and that they should thus at once be driven out of the trade by it, or be gradually forced to quit it, by the slow and successive advances of the machine? The suffering which arises from a quick transition is undoubtedly more intense, but it is also much less permanent than than which results from the slower process; and if the competion is perceived to be perfectly hopeless, the workman will at once set himself to learn a new department of his art. On the other hand, although new machinery causes an increased demand for skill in those who make and repair it, and in those who first superintend its use, yet there are other cases in which it enables children and infirm workmen to execute work that previously required greater skill. In such circumstances, even though the increased demand for the article, produced by its diminished price, should speedily give occupation to all who were before unemployed, yet this very diminution of skill required, would open a wider field for competition amongst the working classes themselves."*

This a tolerably fair statement of the general bearings of the question, but does not take into consideration either the condition of the workmen, or the history of machinery, as applied to manufactures, when put into motion by steam-power.

In the first place, the object of every mechanical contrivance is, to do away with the necessity for human labour, which is at once the most expensive and troublesome agent in the production of manufactured articles. This point has not yet been attained: but already facilities for increased production have been given, which, by the diminution in the first outlay, from the use of steam power and machinery, and by the immense quantity produced, lowering its market value, have rendered it necessary that workmen should individually turn off an increased amount of work, to keep them on a level with the state of things brought about by the two former causes. This at present they are enabled to do; and their earnings have kept on the whole more steady than is generally imagined,† notwith-

* Babbage, Economy of Manufactures, p. 336.

† "The increased production arising from alteration in the machinery, or from improved modes of using it, appears from the following table. A machine, called a "stretcher" in cotton manufacture, worked by one man, produced as follows:—

Year.—Pounds of Cotton spun.—Roving Wages per Score.—Rate of Earning per Week					
		s.	d.	s.	d.
1810	400	1	3½	25	10
1811	600	0	10	25	0
1813	850	0	9	31	10½
1823	1000	0	7½	31	3

standing the astounding fact, as stated by Mr. Marshall, that the same quantity of work is now performed for 1*s.* 10*d.* for which 16*s.* were paid in 1814.

Again, from the simplification in the processes of manufacturers, and whatever power is required being given by steam, to the exclusion of that of man, the family of the manufacturing labourer, from nine years of age and upwards, are all able to earn something; so that when all things are considered, it is probable that, as far as pecuniary mat-

---

" The same man, working at another stretcher, the roving a little finer produced :

Year.—Pounds of Cotton spun.—Roving Wages per Score.—Rate of Earning per Week.					
		s.	*d.*	*s.*	*d.*
1823	900	0	7½	28	1½
1825	1000	0	7	27	6
1827	1000	0	6	30	0
1832	1200	0	6	30	0

" In this instance production has gradually increased, until, at the end of twenty-two years, three times as much work is done as at the commencement, although the manual labour remains the same. The produce of 480 spidles of mule-yarn spinning at different periods, was as follows :—

Year.—Hanks about forty to the pound.—Wages per thousand.			
		s.	*d.*
1806	6,668	9	2
1823	8,000	6	3
1832	10,000	3	8."—

Babbage, p. 338.

The same rate of increased production, and diminution of wages for quantity, has marked the weaving department. In both cases, however, the wages have remained moderately steady.

ters are concerned, he is as favourably placed, as at any former period of his history.

But then, in opposition to these, must be taken into account, their incessant occupation, without a moment's pause or relaxation, bound down by the motions of the engine, which requires their untiring and most assiduous attention. They are indeed its bond-slaves, and it is a severe task-master—no interval is allowed—no voluntary cessation. Many scattered allusions have been made in previous portions of the work on this subject, in connexion with other circumstances influencing their character.

It has been remarked by Mr. Babbage, that " if the competition between machinery and human labour is perceived to be perfectly hopeless, the workman will at once set himself to learn a new department of his art." Were this possible, the necessary consequences of mechanical improvement would signify nothing; but it is impossible, and a reference to his own table on hand-loom weavers will sufficiently show that there are insurmountable difficulties in the way of the conversion of a great body * of operatives from one industrial condition to another.

Whoever is in the habit of visiting the work-

* " It appears that the number of hand-looms in use in England and Scotland in 1830 was about 240,000—nearly the same number existed in 1820. During the whole of this period the wages and employment of hand-loom weavers have been very precarious."—*Babbage,* p. 340.

shops of the machine-makers, and the mills of the great cotton manufacturers from time to time, cannot fail to be struck with the incessant improvements in the application of machinery. These improvements, though they may not enable the master to dismiss any of his hands, prevents the necessity for engaging fresh ones, though he doubles the productive powers of his mill. Thus, in 1806, from the table given above, it appears that 300 men would have been required to turn off the same quantity of work as is now turned off by 100; and this disproportion is constantly increasing. It is here that machinery threatens the manufacturing population.

The rapid growth of the staple branch of manufacture*—the cotton trade, has caused vast immigrations into those districts in which it is principally carried on. The depression in the agricultural counties has pushed these immigrations beyond the demand; the repeated turn-outs have brought sudden accessions of new hands in great numbers—thousands of Irish have deserted their native and miserable homes, in search of employment at the loom; these circumstances, one and all, have brought into the trade a surplus quantity of men, and that at a period when the necessity for them is daily lessening.

---

* The extent to which this branch of trade has already reached is most astonishing, for in 1830 the exported cotton goods were fully one-half of all the exports of the kingdom.

The time too, must come—though it cannot come very quickly, when the trade will reach its maximum. Great Britain will long retain her preeminence as a manufacturing country, in consequence of the progress she has already made, which places her immeasurably in advance of the whole world. Her wealth, the industry of her people, the enterprize of her merchants, the possession of coal, the skill of her artizans—these will long secure her the lead. In retaining this, however, the population engaged in the manufacture will have to suffer greatly. Other nations, in their turn, will devote their energies to the same purpose; and though their present inferiority as to roads, canals, skill, and enterprise, combine to render their competition not very formidable, this will not long continue. Already the United States, France, the Low Countries, and Switzerland, are striding in the same track, and already do something to lessen the value of English manufacture. They are as yet but in their infancy. It is a trade however, which can advance with great rapidity, when energetically taken up; and the period is not very remote, when powerful rivals will dispute the ground with Great Britain, and compel her manufacturers to adopt such measures, and to lower the value of human labour to a level that will enable them to meet upon equal terms, the foreign trader. The surplus hands would readily enable them to lessen the rate of wages; but the certainty that any considerable reduction would be followed by immediate immense losses from turn-outs, extended stoppages, and various other impediments which would be thrown in their way, they prefer the slower process of mechanical improvement, by which, though they may triple the production, but require no new men.*

---

* To show the extent of foreign competition *at present,* the following data are given:—

It appears that in 1821 the exportation of raw cotton from the United States of America to France, and the European continent, was somewhat more than *one-sixth* of the quantity sent to Great Britain. In 1832 it amounted to more than *one-half.*

In 1831 France produced 63,000,000 lbs. of cotton yarn, which will have required a consumption of nearly 80,000,000 lbs. of the raw material. In the same year the *export* of cotton goods from that country amounted to 2,192,240f.

In Switzerland the consumption of cotton in 1831 amounted to 19,000,000 lbs. Only a few years have elapsed since that country received the greatest part of her supplies of yarn from England; she now not only receives very little below seventy shillings, but she has herself become an exporter to other countries.

The manufacture of cotton is rapidly increasing in Prussia throughout the Rhenish provinces, in Silesia, in Saxony, and in both the German and Italian provinces belonging to the crown of Austria.

From information obtained on the spot, or replies to questions which were sent to manufacturers on the continent, it appears that they can compete successfully with England in coarse numbers—some say as high as 40s., others as high as 70s. and 80s. It must be kept in mind the quantity of yarn spun in England above 80s. is very small—about the proportion of 100 or 120 lbs. of coarse for one pound of fine.

The wages paid per week in England and the continent are as follows:—

	England.		Continent.	
For spinners . . .	20s. to	35s. . .	8s. to	10s.
Piecers	4s.	8s. . .	2s.	3s.
Card-room hands . .	6s.	14s. . .	4s.	5s.
Reelers	7s.	13s. . .	4s.	6s.

But these improved applications of mechanism will not end here. Whenever the pressure of foreign or domestic competition becomes more severe, the masters will be necessiated to avail themselves to the utmost of every thing which can assist in lowering the price of their products, and human labour must and will be pushed to the wall. Many great changes will of course take place before this inevitable result is gained, and reductions in wages for quantity will be constantly progress-

---

There is no cotton twist of low numbers now exported from England to France, or Switzerland. Last year there were several orders came for yarns from Germany, restricted to a certain price, but stating that this fixed price was what it could be spun for in that country—these prices were ¼d per pound lower than they could possibly be produced here.

In America the official report of the committee of the New York Convention shows a consumption in 1831 of 77,516,316 lbs. of cotton, which equals the quantity consumed in Great Britain twenty years ago. American fabrics are rapidly superseding those of this country in Mexico, the Brazils, and, in fact, throughout all the markets of the South American continent, and have been brought this year in much larger quantities than heretofore into the ports of the Mediterranean, which can only be accounted for by the success which has attended their previous transactions.

It will also be kept in mind, that two very important items in the cost of a piece of cotton cloth, are raw cotton and food. How much greater, therefore, must the advantages of these rivals, who obtain both these articles free from taxation, pay no poor rates, and negotiate for manufacturing labour free from time bills, &c.

The superiority of machinery, and trade details possessed by Great Britain, are abundantly obvious from the above details. It enables her to compete with manufactures who have the opportunity of paying one-fourth the price for the raw material, although she pays her workmen a fourfold price for his labour.

ing; but the ultimatum is less remote than those interested in it are aware off, for let it be remembered that all mechanical applications, and the moving power derived from the expansive nature of steam, have as yet but arrived at one point in their career, and this point says nothing as to what may be done. There can be no question whatever that many processes for which the human hand is at present indispensible, will very shortly have machines adapted to them, that, if they will not quite displace the workman, will render one man capable of producing, or rather of superintending, the production of quantity now requiring ten or twenty labourers. This is no theoretical opinion, the whole history of the cotton manufacture attests its truth, and collateral proofs are abundant in other branches of manufacture.

It does not follow that improvements in existing machinery, or every new machine, should at once throw out a number of hands. Those, however, who argue that machinery never has that effect, and never will have it, either wilfully delude themselves, or take a very limited and imperfect view of the subject. It must have one of two effects, the objects of every change, improvement, and addition, being to lessen the amount of labour required for production, and this must be either to render fewer workmen necessary to produce a given quantity of manufacture, or so far lower the price of the manufacture, as at once to increase the demand for it so considerably as to absorb the same number of men as are already engaged in it. In many instances, in fact, generally the latter, has been the case hitherto, and would, perhaps, continue to be so, were Great Britain entirely to monopolize manufactures. But this cannot be, and as it has been before stated, the maximium must be attained. All these improvements having therefore one end, all tending to the same point; namely, the cheapening of labour. The time must come, when its value will be so small, as to make it nearly worthless to the possessor.

The consequences of this it would be rash to predicate. There is, however, no reason to suppose that some other great national branch of industry may not develop itself as the cotton trade has done. Steam, which has effected so much, may do vastly more in altering or increasing other divisions of manufacture; and there are other material agents, which may not improbably be brought to bear upon production, and so diminish its value beyond any thing at present calculated upon, as to require all the human aid that can be found foı it at a remunerating price.

A crisis has, however, partly arisen, and it behoves every man, interested in the welfare of his country, to examine into the present means for lessening its inevitable evils. An ingenious and eloquent writer in the Quarterly Review has said, "So far are we from regarding the increased use of machinery as an evil which requires to be checked, that we hail every such application of the discoveries of science as another step in the steady course by which the Author of Nature pushes forward the improvement of the human race. In our opinion, instead of being an evil to be deprecated, and if possible counteracted and repressed, the application of machinery, as a substitute for labour, serves to disengage a large number of human beings from manufacturing toil, in order that they may be employed in perfecting and extending our tillage, thereby at once increasing their own happiness and the resources of the empire."

He goes on to say, "We have arrived at a great and most important crisis of social arrangement. We are embarrassed with a superfluity of human labour, of animal machines, which cannot be absorbed in manufacturing operations. What is to be done with this superfluous, or rather disposable fund of human physical power: shall these men be compelled to eke out a miserable existence, with half employment and scanty wages? or shall they be thrown upon their respective parishes for eleemosynary relief?"

In speaking of the crisis which led to the introduction of poor laws, and laid the foundation for some extension of manufactures, and that had resulted from excess of agricultural labourers, and the breaking up of the monastic revolutions, he remarks:—

"The extent to which the employment of machinery has been pushed, as a substitute for human labour, has at length brought on a new crisis: it is one essentially different from that which presented itself to the statesmen of the sixteenth century, and which appears to demand a different remedy. Then the agricultural population had become too numerous, whilst a large proportion of the surplus produce of the English land was exported in exchange for wrought commodities:—now, the difficulty is of a totally different origin and kind. So far are our manufactures from requiring an increased supply of hands, that they overflow with workmen, for whose industry there is no profitable demand. The employment of machinery not only stops up the gap through which the surplus of our agricultural population had been used to make its way into manufactories, but it has likewise thrown out of employment a considerable portion of the hands which had been previously occupied in the fabrication of wrought commodities. From both these sources, a number of unemployed hands accumulate: the gradual increase of population produces a surplus of labourers, who cannot find profitable employment in the tillage of our old lands; and to this surplus is daily added a crowd of workmen, whom the extension and improvement of machinery disengages from manufactories."

But these do not constitute all the elements of the crisis which is developing itself. Agriculture is undergoing a transition as great, and almost as remarkable as manufacture—and these are progressing, step by step, to one and to the same end.

Mechanical contrivances for lessening human labour, are sought for with as great avidity in the one case as in the other; and in a single instance, —that of the use of a peculiarly constructed plough for hoeing up potatoes, one man and one horse get through as much work, as would, a few years ago, have required at least thirty labourers, and perform the task much more completely and efficiently: and this is only one solitary example. The current of prejudice has, hitherto, run strongly against the use of machines for farming purposes, and has been kept up by the limited intelligence existing amongst farmers as a class. The depression which has been gradually but steadily creeping over them, has, at length, succeeded in breaking up the cottier and small farm system, and has thrown agriculture into a new shape and into new hands; which has brought all the force of this epoch of mechanism to bear upon it. The same causes are at work, therefore, upon the two great divisions of national industry, and their effects have even been more severely felt by the agricultural than the manufacturing labourer; and have, in a great measure, already pauperized the whole body, and nearly extinguished the peasantry, as a moral and independent class of the community. The present crisis involves not one particular portion of labourers, or one particular branch of trade, but the interests and future welfare of all, are intimately connected with it. The time, indeed, appears rapidly approaching, when the people, emphatically so called, and which have hitherto been considered the sinews of a nations strength, will be even worse than useless. When the manufactories will be filled with machinery, impelled by steam, so admirably constructed, as to perform nearly all the processes required in them, and when land will be tilled by the same means. Neither are these visionary anticipations; and these include but a fraction of the mighty alterations to which the next century will give birth. Well then, may the question be asked—what is to be done? Great calamities must be suffered. No transition so universal, so extensive can be operated without immense present sacrifices; but upon what class, or what division of property or industry these must be more especially inflicted, it is impossible clearly to indicate. Much should be done—and done vigourously and resolutely. Like other great revolutions in the social arrangement of kingdoms, it is to be feared that the explosion will be permitted to take place, undirected by the guiding hand of any patriotic and sagacious spirit, and its fragments be again huddled together in hurry and confusion; and finally to undergo a series of painful gradations, from which the imagination turns with sickening terror.

CHAPTER XII.

TRUCK AND COTTAGE SYSTEMS.

Truck System, what—Evil Influence of—Consequences of Truck Payment—Extent of Truck System and its Mischief—Cottage System, what Reasons for building—Advantage to Men and Masters by bringing them near their Work—Benefit of Isolation to Workmen—Evils of crowding together—Beer Act—Its evil Effects—Mode of evading Excise Laws—Whistling or Straw Shops for the Sale of Beer described—Causes of which led to their Establishment—Injury to Public-house Property, and to the Character of Public-house Keepers—Pecuniary Advantage of Cottage-building to the Master—Its enormous Amount—Its Perpetuation of Trucking—Advantages possessed by Masters over common Holders of Cottage Property—No Risk as to Rent, &c.—Unfairness of Comparison—Injury to the Character of the Masters—Duties of Masters—These described—Truck and Cottage Systems departures from these—Profits derived from them—Injustice to the Men—Importance of a clear Understanding of the relative Duties of Masters and Men—Moral Evils—Slavery—Proper Conduct of Masters—Proper Conduct of Men—Comfort not solely dependant upon Amount of Earnings—Upon what dependant, &c. &c.

The impossibility of the legislature effectually interfering between the master and the labourer, is sufficiently proved by the failure of the laws with respect to the truck system. It is true they have in some degree checked the evil, or rather have compelled the master to mask their dealings so as to evade their penal operations. The truck system as it is termed, is the payment of the labourer not in money, but goods, thus converting the master into a shop-keeper or a retail dealer.

This is a species of combination, which is injurious alike to the character of the master and his men, and one the abolition of which it would be very desirable to see for both their sakes. It is unjust to the labourer in many ways—it assists in the ruin of his feelings of independance and free agency, and keeps down his provident wishes,—in as much as he has no opportunity of being provident.

Next to the evil influence of combinations of workmen upon the social and moral character, is the truck system. It is one which deserves severe condemnation, and which the masters will do well to reflect upon, for it is founded on injustice, and enables them to oppress the men, whilst at the same time it deprives them of all means of resistance.

It has been urged by some writers,* that the

* "When the number of workmen living upon the same spot is large, it may be thought desirable that they should unite together, and have an agent to purchase by wholesale those articles which are most in demand, such as tea, sugar, bacon, &c., and to retail them at prices which will just repay the wholesale cost, together with the expense of the agent who conducts their sale. If this be managed wholly by a committee of workmen, aided perhaps by advice from the master, and if the agent is paid in such a manner as to have himself an interest

men should unite and establish shops for their own supply of the necessaries or comforts of life, such as tea, coffee, sugar, soap, &c. &c. There is no question that the retail dealer, the petty shop-keeper, gains a very large profit by his sales, partly by overcharging, and partly by adulterating his articles, and that so far the men would improve their situation if they could do away with this. A little acquaintance with their characters and feelings would dissipate any notion of preconceived advantages, which they would derive from an establishment of a joint stock concern, in the majority of instances, and besides it is impracticable from many causes. They will suffer the least from an adherence to the old plan both in a pecuniary and in a moral point of view—and the retail dealer will be by simple competition brought to something like a moderate demand for the employment of his capital.

The case, however, is widely different with regard to the master. They have indeed urged that the men are still free to purchase at their shop or elsewhere as best suits their convenience. The statement is fallacious and unworthy consideration—the men are not, cannot be free agents in the matter. Abundant means are in the hands of

---

in procuring good and reasonable articles, it may be a benefit to the workman."—*Babbage*, p. 308.

There are innumerable difficulties in the way of Mr. Babbage's scheme, advantageous as it appears at first sight, and in practice it would be found totally useless.

the master to compel them by indirect measures to confine themselves to his shop, and in very numerous instances no such reserve is shewn, but the men are presented with a ticket to nearly the full amount of their wages, and this is alone negotiable at these shops. It has happened repeatedly that the workman wanting a few shillings for some other purpose, has consented to take a sum considerably less than was his right rather than have articles which he either did not want, or that he wanted something else more indispensably and immediately—and this was his only mode of procuring the needful sum at an immense sacrifice.

If the masters object was the advantage of the labourer, and if the labourer could be induced to give him credit that it was so, and if the master employed a portion of his capital to the purchase of wholesome goods, and retailed them at a lower price than the common shopkeeper, which he could well afford to do, and still have a fair return, some of the objections against the system would fall to the ground. But it is just the reverse—the price charged by the master is invariably as high and in many instances much higher than his humble neighbour, and his goods very generally inferior, and retailed out so arbitrarily as to increase the other mischiefs. Many masters in all branches of manufactures are in the habit of killing cows, sheep, pigs, &c., very extensively, and distributing the meat amongst their men, at a price universally above the market average; many are dealers in coal, and in short every commodity required by the operative, some or other of these masters deal in.*

The evils of this system are so apparent, and its injustice so gross and glaring, that nothing but a very low estimate of morality can be possessed by those masters who are engaged in it. They drive all competitors at once from the field, ruining numbers of decent retailers and forcing them a step downwards in the social scale, by converting them into labourers, and obliging their families, who were most probably living at home in domestic privacy, and preparing themselves for decent members of society, into the mill,—a situation certain to do away with many of those probabilities. They thus prevent any middle class betwixt themselves and their operatives from springing into permanent existence—a class to which the industrious, economical, and well-disposed workman might aspire—they smother all disposition for social amendment, and thus both directly and indirectly keep up and increase the improvidence and indifference of the labourer. The establishment of a middle class between them would be exceed-

* "If the manufacturer kept this shop merely for the purpose of securing good articles, at fair prices to his workmen, and if he offered no inducement to them to purchase at his shop, except the superior cheapness of his articles, it would certainly be advantageous to the men. But, unfortunately, this is not always the case.—*Babbage,* p. 309.

ingly beneficial to the character of both,—to the man by shewing within his reach a condition in life elevated above that which he now holds—to the master by establishing a barrier which would separate him for troublesomely and vexatiously interfering with the details of his inferior.

Another system rapidly progressing, and very much resembling the truck system, of which in fact it may be viewed as an off set, is the cottage system. A master having in his employ several hundred hands, whose habitations are scattered at considerable distances from each other and from the mill, erects ranges of cottages in its immediate vicinity, forming a part indeed of itself. These cottages are probably better built, more commodious, in every respect more comfortable than those which the labourers previously inhabited; by bringing them nearer their work, they escape exposure to the weather in their progress to and from their work; they have also an opportunity of securing longer hours of rest by the saving of time which was occupied by these progresses; and yet the system is a bad one, and one decidedly injurious to the men. This system of cottage building, it is true, is not universal in the towns where masses of labourers are already collected, and when considerable expence would attend upon it, it is very little practised. It is chiefly in the out-districts, and it is a striking proof of the advantages to the master, that the great manufacturers are gradually creeping to the outskirts or into localities a few miles from the great towns.

The extension and influence of this system may be very distinctly seen in the now populous township of Hyde and Newton, Duckenfield, &c. about seven miles from Manchester, and three from Ashton.

The population of these districts in 1801, scarcely amounted to three thousand, whilst in 1830 it had increased to twenty-six thousand nearly, a rate of increase unequalled. The rapid growth of such a population in this neighbourhood, is of course owing to the concentration of a number of manufactories here, favoured as their position is by an abundant supply of coal and excellent means of transport. This population which is the resident one, by no means indicates the number of hands employed, on the contrary, great numbers are furnished by the surrounding townships.

An examination of the habitations of this population would show how very large a proportion is in the hands of the manufacturers, either as owners, or general tenants. It may be urged, that the masters in consequence of the increase in the number of miles, and the consequent difficulty of collecting hands from a distance, have been forced to build, nobody else indeed being found who would speculate upon such property. This is very true, and quite unanswerable. They have been forced to build, and are still building; but it does not lessen the evils attendant upon it.

The demoralizing effects of crowding together promiscuously the labourers in factories, have in large towns been materially aided by the nature of their habits and the vices which large towns afford facilities for pursuing. So long as the operative, after having completed his day's toil, was seperated from his fellows, and had to seek his isolated cottage, some of these evils were avoided. It freed him for a time at least from bad example, removed him from immediate contact with the gin-shop or tavern, collected his family generally under one roof, and gave him an opportunity for cultivating his social affections, and keeping down the overwhelming force of demoralization, which too often destroyed the town labourer. By affording him the cheering influences of a pure atmosphere, the sights and sounds of natural objects, the out-district labourer presented for a time some moral and physical traits, which made him superior to the town operative.

The Beer Act has unhappily destroyed one of these advantages. The race of keepers of public-houses in the rural districts, was, generally speaking, one something superior to his immediate neighbours *—a small farmer or shopkeeper, and reputable householder. Since that period, however, a great change has been working in the

* Vide Introductory sketch.

licensed victuallers. The small sum required for a license, threw it open to a labourer, and the idle and profligate took immediate advantage of it. Numbers of beer-houses were speedily seen occupying sites, where as a matter of trade, very little advantage could be derived from them. Every way-side, every clump of cottages, shewed, as it was termed here, its " Tom and Jerry shop," a term singularly appropriate. It cannot be denied, that prior to the introduction of these, an organized system for evading the excise laws was developing itself, in a very singular, and yet in a very effective form. The public-houses at this period, (prior to the passing of the Beer Act) it has been said, was a house of good report, where, though immorality and drunkenness would at times shew themselves, still they were checked and kept in some sort of discipline. The bringing together numbers of the young of both sexes by the factory system, generated irregularities, which had no adequate field for their display; drinking became more habitual, and the more it was indulged, the more it required indulgence. To meet this demand of demoralization, a number of houses were opened in the vicinity of the manufactories, termed " whistling-shops" or " straw-houses," for the sale of beer, which however, as the parties were unlicensed, subjected them to a penalty. To evade this, no beer was sold, but abundance was given. The mode in which this transfer was managed, to the advantage and compensation of the giver, was, that a straw was the matter in barter; the drinker bought a bit of straw for the price of a pint of beer, and this pint was given in with the bargain. Odd as this may seem, it was carried on very extensiuely, and exists yet to some degree, though it has been displaced partially by the opening of beer-houses, which cost little, and free their owners or occupiers from the risk of punishment. By lowering the value of public-house property, and by lowering in an equal degree the character of the holders of these houses, great moral injury has been done, without one single advantage to the poor man to compensate for it. It was a law based upon most erroneous and mistaken principles—holding out to the labourer an apparent addition to his comfort, but which has signally and unhappily failed.

The advantages possessed by the operative living in a detached situation, were counterbalanced in some degree, by the increased labour, which, if he lived two or three miles away from the mill—a very common distance, he necessarily underwent, and by consequent exposures from a heated atmosphere to a natural temperature in unfavourahle weather, and by the necessity for early rising and late going to bed. When however, the distance was more limited, he derived benefit from the change of exercise, night and morning, but was prevented going home to dinner. The advantages of most importance were morals, and these he loses by being brought under the cottage system.

The masters, many of whom have 80, 100, 200, or more of those cottages surrounding, or in the immediate neighbourhood of their mills, are immense gainers by the arrangement. The cost of building a range of houses such as these, say 100, will not, upon the most liberal average, be more than 5,000*l.*; indeed it may be estimated at vastly less; for the generality of them – taking the very best, 50*l.* per house will over and above cover the outlay. Now for the outlay of 5,000*l.*, the capitalist draws an annual income of 800*l.*, or 13½ per cent., and completely covers himself in little more than six years. This profitable return is burdened with no drawback, no rent is lost, every pay-night it is deducted from the wages. One cannot wonder therefore, at the universal practice of manufacturers building cottages, favourably placed as they are. Besides this disproportionate rate of interest, he derives other great advantages from thus congregating his men under his immediate controul. In the first place, it enables him without difficulty to perpetuate the truck system: for he invariably builds two or three shops, and houses calculated for taverns or beer-houses; and if he has no direct dealing in them, he abstracts an equivalent rent; and his hands are expected – and do spend a considerable portion of their earnings there, so that they suffer precisely the same injury as if their master was the retailer; and they know it. Again, he derives benefit in another way: it enables him to shorten the hours allowed for meals—to begin earlier in the morning, to continue later at night—and this too with the concurrence of the hands, who often do not understand the difference sufficiently between mill labour and walking on the road in the open air; and yield up an hour at least per diem, which is added to the profits of the cottage building.

It is argued by those who advocate the system, that there are many circumstances favourable to the men in it—that generally a school-house is built, and every disposition shown by the master, to forward the comfort and domestic enjoyments of the men. That the houses are commodious, clean, whitewashed, and in every respect vastly superior to the habitations for a similar class of labourers in the town. Any one who will take the trouble to inspect them, will at once acknowledge the truth of these assertions; but in opposition, it must be born in mind, that all these advantages in the domestic condition of the hands, are sources of large pecuniary profit to the master, at their expense. It is said the rent, 3s. per week, is not more than the tenant would have to pay for a cottage *as near* the mill, to any other landlord. This is not denied—but the argument is unfair. The landlord who has cottage poperty, charges probably the same per-centage in towns and populous neighbourhoods: but he is liable to loss of rent continually—to have his houses empty for weeks and months together—to have his tenants leaving his houses, almost in a state of dilapidation. &c. &c.

so that, although he has the same nominal rent, and is so far on a par with the manufacturer, he seldom calculates upon receiving, or ever does receive, on an average more than six per cent., at the utmost—rarely indeed so much. It is, therefore, obviously and clearly a misnomer, to say that the one derives the same profit as the other; whereas the manufacturer receives without risk or trouble, or even collecting, 13½ per cent: whilst the common capitalist has very considerable difficulty in securing his five or six per cent; and that has to be waited for and got together with incessant exertion.

It would be highly gratifying if the masters, in extending this system, which might have been the agent of much good to the men—had not shown their rapacity so plainly. Had they contented themselves with fair interest, and accommodated their hands with these cottages, it would have been well—it would have been gratifying to every one, desirous of seeing a change in the intercourse between them. They might have, indeed, contented themselves with even a less amount of interest than that paid by building property in general, and still have been gainers, by having the opportunity of lengthening in a very limited degree, the hours of labour, and by many other incidental advantages which they gain, by bringing the men in such immediate contact with the mill. But when this system is made a source of great revenue—and when to this it superadds the truck system, with all its vices, it can neither improve the condition of the labourer, nor add to the honourable name of the master.

The truck system and the cottage system are both departures from the proper track and duties of the master—both bring him into collision with his men, upon points on which he ought never to interfere. His duties are so strongly marked that he should never suffer a question which can, for a moment, in any way make him diverge from their proper performance, to be entertained. He has a large capital employed, and for the purpose of turning it to profitable account, he requires the assistance of a certain number of labourers, the expense of the hiring of which is added to the cost of the peculiar manufacture he brings into the market. This payment of labour should be of that extent, and no farther, as will enable him to compete with other traders; and should be greater and smaller as the article he produces sells at greater or less price than all the expenses of its production, such as raw material, building, machinery, labourers, &c. leaving to the speculator, or producer, a fair profit for time and capital expended. This constitutes the profit of the manufacturer; and this done, he is bound to pay his labourers in money, their part of the expense of production—namely for their labour. But the truck system and the cottage system go beyond this. The master pays or gives an acknowledgment to the men for their quota of profit, which he declares to be that which the present state of the market enables him to make; but he is not content with taking his share in the first instance, for he taxes their wages, and though nominally not reducing them, subtracts twelve per cent. from their gross amount

One mode in which these systems are prejudicial to the interests of the workmen is, that it gives the masters an opportunity of lowering their wages in times of trade depression, or at their own caprice, and that too in an underhand way, which they can with difficulty resist. Sixpence a week in addition to the rent of their houses, and a halfpenny or farthing a pound or yard on the articles of their consumption, are innovations not so great as to arouse their attention or indignation; but it gives a large sum to the masters, and the temptation thus held out is too great to be resisted. Hence a series of frauds upon the men—for frauds they decidedly are, of the most injurious tendency.

A manufacturer employs 600 hands, the entire sum of whose wages for six months, at an average of ten shillings per head, is 7,800*l.* This he pays as the value of their labour, after having allowed for every other item, and his own profit. If his men are subjected to the truck and cottage system (and remember the first amount, which he declares is one left to his own conscience, and is not improbably much below its fair value), he levies upon their capital a tax of nearly 500*l.*, and this over and above the customary profits made upon the articles with which he supplies them, namely, houses and goods, and all this without risk.

It has been previously urged that combinations of the masters are needful and desirable when founded upon *justice* and *equity*, but this monstrous perversion of both cannot be sufficiently condemned.

The capabilities which the labouring man has to improve his condition under the numerous disadvantages to which he is subjected, cannot receive a more striking exemplification than from the fact, that the hands who, in addition to their other disadvantages, are farther subjected to those of these two systems, are in many respeets decidedly improved in their social and moral organization, and nothing can shew more clearly what might be done by masters, than what interest based upon injustice has led them to.

This injustice comes with double force upon the operative, as it injures him indirectly, and in a way which he does not understand, and in a way, too, that he cannot resist. He perhaps does not feel its operation so onerous as in reality it is, and struggles on complaining against a rate of wages, which is hardly adequate for his wants and expectations—combining, too, against them, paying a further sum on this account, while the actual cause of his misfortunes lies shrouded and covered by so specious a show, that he never dreams of imputing them to it.

All the relations between master and man should

be clearly understood—they are simple when freed from suspicions on the one hand, and grasping avarice on the other. Their compact is one for the interest of both, if they will properly cultivate it; and too much tenacity in clinging to supposed rights or immunities is injurious to all parties. The labourer has a positive right to the market value of his labour, whether more or less, whether sufficient or insufficient to support him; and the master ought to give such explanation as will satisfy him that he has such value, and neither should go beyond this. The labourer should not interfere with the regulations of his master, nor should the master interfere with the free agency of the labourer. Each have their proper functions, and like the members of the human body in the well-known fable, if they quarrel, they will most assuredly suffer.

The moral evils of the truck and cottage systems are analogous. By reducing the labourer to a mere machine—by destroying his personal independence—by cutting off his claim to self-respect—they degrade him to a condition of mere slavery, compared to which the West Indian slave may indeed congratulate himself on his good fortune, for his is a state to be a thousand-fold desired in preference. The name of slave is a bye-word for conjuring up a frightful catalogue of miseries and degradations; but the slave has no master like the steam engine, none which requires such assiduous, such untiring attention. He is not compelled by this inexorable tyrant to toil for twelve or fourteen hours uninteruptedly; and be it remembered that the factory labourer works in an atmosphere of equal temperature with the slave—and shall a name make such mighty difference? When the moral and physical conditions of slavery are equally in force in both instances? for so it is when the master takes under his controul the hard won wages of his labourers, and again converts them to his own use, leaving them to suffer any degree of hardship which will not drive them to desperation.

They both necessarily render the men improvident and reckless—destroy all forethought—paralyze every better feeling within them—and thus brutalize and degrade them in exact ratio as they are made dependant and hopeless of amelioration or amendment.

It is hardly possible that the masters will ever sufficiently comprehend, that to make money is an evil, or that they will possess such honourable feelings as will lead them to sacrifice a positive advantage over the men, when it is entirely in their power. Even were it so—even were masters disposed to consult these things, it is not likely the world, or at all events their men, would give them credit for it. Let them, therefore, wash their hands of all temptation; let them shew, by not having, that they do not wish to have unfair advantages; and though they may lose something for a time, they will eventually be large gainers; for it will be the first great step towards improving the moral and social condition of the mass of the labourers—which, by rendering them economical, introducing household virtues, withdrawing them from unnatural stimuli, inducing them to seek their support from solid and wholesome diet, will enable them to do all these things at a rate of wages greatly inferior to what they now contrive to squander, and yet starve—to labour incessantly, yet seem to labour for a pittance unequal to supply their wants. It is indeed difficult to say what is the minimum which a provident man and his family can subsist upon: one thing is obvious to every man who will inquire, that an amount of earning which appears to one family totally inadequate to live even decently with, when all about is squalid beggary, and everything shews the extremity of want—yet that another family, equally, if not more numerous, will, with the same amount of wages, appear clean, well fed, well clothed, cheerful, and happy. It is not in the amount of a man's earning, so much as to his capability of turning this to the best account. Great wealth is no safeguard against occasional distress, if not properly economized and managed; neither are liberal wages any reason why those earning them should not be pauperized. The same family, to the present writer's positve knowledge, has at one period earned £5 per week. Trade has become depressed—prices have fallen—and in a year or two their earnings have been reduced to £2 10*s*. But were they poorer? By no means. Riches and poverty are merely terms of comparison. Were they worse dressed? No!

Worse fed? No! Did they become burdensome to the parish? No! How then? They learnt to reduce their superfluous expenditure; to cut off all unnecessary sources of outlay; to live more regularly,—and they were, in every respect, as well to do in the latter period as in the former. Thus it will be, if the manufacturing population can be roused from their present evil habits; if they will make home what it should be; if the earnings of families are brought into a joint fund; if domestic discipline is re-established; and if they become good husbands, fathers, brothers, and wives; if they will limit the demands of their appetites to their means of supply; throw aside ruinous combinations; become peaceable, orderly, and well-behaved citizens; devote their earnings to legitimate purposes, and eschew alike evil and folly, they have within their power—means to be comfortable, happy, and perfectly independent; and to defer, or perhaps do away with the chance of the frightful consummation of their misfortunes, which their present modes of life is so rapidly and so fatally accelerating.

THE END.

LONDON:
BAYLIS AND LEIGHTON, JOHNSON'S-COURT, FLEET-STREET.

THE EVOLUTION OF CAPITALISM

Allen, Zachariah. **The Practical Tourist,** Or Sketches of the State of the Useful Arts, and of Society, Scenery, &c. &c. in Great-Britain, France and Holland. Providence, R.I., 1832. Two volumes in one.

Bridge, James Howard. **The Inside History of the Carnegie Steel Company:** A Romance of Millions. New York, 1903.

Brodrick, J[ames]. **The Economic Morals of the Jesuits:** An Answer to Dr. H. M. Robertson. London, 1934.

Burlamaqui, J[ean-] J[acques]. **The Principles of Natural and Politic Law.** Cambridge, Mass., 1807. Two volumes in one.

Capitalism and Fascism: Three Right-Wing Tracts, 1937-1941. New York, 1972.

Corey, Lewis. **The Decline of American Capitalism.** New York, 1934.

[Court, Pieter de la]. **The True Interest and Political Maxims, of the Republic of Holland.** Written by that Great Statesman and Patriot, John de Witt. To which is prefixed, (never before printed) Historical Memoirs of the Illustrious Brothers Cornelius and John de Witt, by John Campbell. London, 1746.

Dos Passos, John R. **Commercial Trusts:** The Growth and Rights of Aggregated Capital. An Argument Delivered Before the Industrial Commission at Washington, D.C., December 12, 1899. New York, 1901.

Fanfani, Amintore. **Catholicism, Protestantism and Capitalism.** London, 1935.

Gaskell, P[eter]. **The Manufacturing Population of England:** Its Moral, Social, and Physical Conditions, and the Changes Which Have Arisen From the Use of Steam Machinery; With an Examination of Infant Labour. London, 1833.

Göhre, Paul. **Three Months in a Workshop:** A Practical Study. London, 1895.

Greeley, Horace. **Essays Designed to Elucidate the Science of Political Economy,** While Serving to Explain and Defend the Policy of Protection to Home Industry, As a System of National Cooperation for the Elevation of Labor. Boston, 1870.

Grotius, Hugo. **The Freedom of the Seas,** Or, The Right Which Belongs to the Dutch to Take Part in the East Indian Trade. Translated with Revision of the Latin Text of 1633 by Ralph Van Deman Magoffin. New York, 1916.

Hadley, Arthur Twining. **Economics:** An Account of the Relations Between Private Property and Public Welfare. New York, 1896.

Knight, Charles. **Capital and Labour;** Including *The Results of Machinery*. London, 1845.

de Malynes, Gerrard. **Englands View, in the Unmasking of Two Paradoxes:** With a Replication unto the Answer of Maister John Bodine. London, 1603. New Introduction by Mark Silk.

Marquand, H. A. **The Dynamics of Industrial Combination.** London, 1931.

Mercantilist Views of Trade and Monopoly: Four Essays, 1645-1720. New York, 1972.

Morrison, C[harles]. **An Essay on the Relations Between Labour and Capital.** London, 1854.

Nicholson, J. Shield. **The Effects of Machinery on Wages.** London, 1892.

One Hundred Years' Progress of the United States: With an Appendix Entitled Marvels That Our Grandchildren Will See; or, One Hundred Years' Progress in the Future. By Eminent Literary Men, Who Have Made the Subjects on Which They Have Written Their Special Study. Hartford, Conn., 1870.

The Poetry of Industry: Two Literary Reactions to the Industrial Revolution, 1755/1757. New York, 1972.

Pre-Capitalist Economic Thought: Three Modern Interpretations. New York, 1972.

Promoting Prosperity: Two Eighteenth Century Tracts. New York, 1972.

Proudhon, P[ierre-] J[oseph]. **System of Economical Contradictions:** Or, The Philosophy of Misery. (Reprinted from *The Works of P. J. Proudhon,* Vol. IV, Part I.) Translated by Benj. R. Tucker. Boston, 1888.

Religious Attitudes Toward Usury: Two Early Polemics. New York, 1972.

Roscher, William. **Principles of Political Economy.** New York, 1878. Two volumes in one.

Scoville, Warren C. **Revolution in Glassmaking:** Entrepreneurship and Technological Change in the American Industry, 1880-1920. Cambridge, Mass., 1948.

Selden, John. **Of the Dominion, Or, Ownership of the Sea.** Written at First in Latin, and Entituled *Mare Clausum.* Translated by Marchamont Nedham. London, 1652.

Senior, Nassau W. **Industrial Efficiency and Social Economy.** Original Manuscript Arranged and Edited by S. Leon Levy. New York, 1928. Revised Preface by S. Leon Levy. Two volumes in one.

Spann, Othmar. **The History of Economics.** Translated from the 19th German Edition by Eden and Cedar Paul. New York, 1930.

The Usury Debate After Adam Smith: Two Nineteenth Century Essays. New York, 1972. New Introduction by Mark Silk.

The Usury Debate in the Seventeenth Century: Three Arguments. New York, 1972.

Varga, E[ugen]. **Twentieth Century Capitalism.** Translated from the Russian by George H. Hanna. Moscow, [1964].

Young, Arthur. **Arthur Young on Industry and Economics:** Being Excerpts from Arthur Young's Observations on the State of Manufactures and His Economic Opinions on Problems Related to Contemporary Industry in England. Arranged by Elizabeth Pinney Hunt. Bryn Mawr, Pa., 1926.